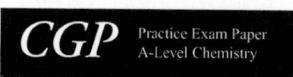

AQA A-Level Chemistry

Practice Paper 1 — Inorganic and Physical Chemistry

Time allowed: 2 hours

Centre name					
Centre number					
Candidate number					

Surname	
Other names	
Candidate signature	

In addition to this paper you should have:
- A Data Booklet (including a Periodic Table)
- A ruler
- A calculator

Instructions to candidates

- Use black ink or ball-point pen.
- Write your name and other details in the spaces provided above.
- Answer all questions.
- Answer the questions in the spaces provided.
- Show clearly how you worked out your answers.
- Cross out any work you do not want to be marked.

Information for candidates

- There are 105 marks available for this paper.
- The marks available are given in brackets at the end of each question.

For examiner's use

Q	Mark	Q	Mark
1		7	
2		8	
3		9	
4		10	
5		11	
6			
Total			

Answer ALL the questions.

Write your answers in the spaces provided.

1 This question is about the structure and bonding of fluorine compounds.

1.1 Use the electron pair repulsion theory to identify the shape
and bond angle of the AlF_6^{3-} ion. Explain your answer.

Shape _____

Bond angle _____

Explanation _____

_____

_____

_____

_____

[4 marks]

1.2 Fluorine and hydrogen fluoride are both examples of simple covalent molecules.
Fluorine has a boiling point of −188 °C and hydrogen fluoride has a boiling point of 19.5 °C.

Explain, in terms of intermolecular forces, why hydrogen fluoride has a much higher
boiling point than fluorine.

_____

_____

_____

_____

_____

_____

_____

_____

[3 marks]

1.3 Manganese(IV) hexafluoride $[MnF_6]^{2-}$ is a complex ion made up of a central manganese atom surrounded by six fluoride ligands.

State the type of bond formed between the central manganese atom and each of the six fluoride ions. Describe how these bonds are formed in $[MnF_6]^{2-}$.

_____

_____

_____

[2 marks]

1.4 Potassium fluoride is a white crystalline solid which has a high melting point and dissolves in water to form a colourless ionic solution.

Describe the electrical conductivity of potassium fluoride, both when solid and in solution. Explain your answer in terms of structure and bonding.

_____

_____

_____

_____

_____

_____

_____

_____

[4 marks]

Turn over for the next question

Turn over ▶

2 A student was investigating the enthalpy change of the exothermic displacement reaction that occurs when powdered zinc is added to 25 cm³ of a copper(II) sulfate solution.

2.1 Write the overall ionic equation for the reaction between zinc metal and copper(II) sulfate solution. Do **not** include any spectator ions.

_____

[1 mark]

2.2 The student measured the temperature of the copper(II) sulfate solution every minute for three minutes. At the fourth minute they added the powdered zinc to the copper(II) sulfate solution but did not record the temperature. The student then recorded the temperature of the reaction mixture at the fifth minute and for every minute thereafter until a total of ten minutes had elapsed.

Describe how the student should process their results to determine the maximum temperature change in this reaction.

_____

_____

_____

_____

_____

_____

_____

_____

_____

[4 marks]

2.3 Another student was investigating the enthalpy change when sodium carbonate is dissolved in water. Leave blank

$$Na_2CO_{3(s)} \rightarrow 2Na^+_{(aq)} + CO_3{}^{2-}_{(aq)}$$

The student dissolved 4.10 g of sodium carbonate in 50.0 g of water in a polystyrene cup. After processing their results, the student determined that the enthalpy change per mole of sodium carbonate was −26.5 kJ mol⁻¹.

The initial temperature of the solution was 21.4 °C.
Calculate the maximum temperature reached by the solution during this experiment. You may assume that only the water changes in temperature and the specific heat capacity of water is 4.18 J K⁻¹ g⁻¹.

Maximum temperature _____ °C

[4 marks]

Turn over for the next question

A-Level Chemistry / AQA / Paper 1

© CGP 2020 — copying more than 5% of this paper is not permitted

3 Time of flight (TOF) mass spectrometry can be used to identify elements from their relative atomic mass.

A sample of element **X** was ionised using electron impact ionisation in a TOF mass spectrometer. The spectrum produced showed that there were four isotopes of element **X** present in the sample. The masses and abundances of the four isotopes in the sample are shown in **Table 1**.

Table 1

Isotope	Abundance (%)
^{84}X	0.56
^{86}X	9.86
^{87}X	7.00
^{88}X	82.58

3.1 Calculate the relative atomic mass of element **X** in this sample.
Give your answer to one decimal place.

Relative atomic mass _____

[2 marks]

3.2 Deduce the identity of element **X**.

_____

[1 mark]

3.3 Each $^{86}X^+$ ion travelled through the flight tube of the TOF mass spectrometer with a velocity of 82200 m s^{-1}. The kinetic energy of an ion can be calculated using the equation:

$$KE = \frac{1}{2}mv^2$$

where KE = kinetic energy (J), m = mass (kg) and v = velocity (m s^{-1}).

Calculate the kinetic energy of a $^{86}X^+$ ion in this flight tube.
Give your answer to three significant figures.
The Avogadro constant is 6.022×10^{23} mol^{-1}.

Kinetic energy _____ J

[2 marks]

Turn over for the next question

Turn over ▶

4 This question is about the Period 3 elements and their oxides.

4.1 MgO, SiO_2 and P_4O_{10} are all Period 3 oxides. Explain why the melting point of SiO_2 is higher than that of P_4O_{10} but lower than that of MgO. Give your answer in terms of structure and bonding.

_____

_____

_____

_____

_____

_____

_____

_____

_____

_____

[5 marks]

4.2 **Table 2** shows some Period 3 elements in different orders.
Which row in **Table 2** shows the elements in the correct order from smallest to largest atomic radius? Tick **one** box.

Table 2

A	S < Cl < P < Al < Na	☐
B	Na < Al < P < S < Cl	☐
C	Al < Na < S < P < Cl	☐
D	Cl < S < P < Al < Na	☐

[1 mark]

4.3 Write an equation for the reaction that occurs when aluminium oxide reacts with hydrochloric acid.

_____

[1 mark]

5   This question is about Group 2 metals and their compounds.

5.1   Give the equation, including state symbols, for the first ionisation energy of calcium.

_____

[1 mark]

5.2   State and explain the trend in atomic radius down Group 2 from magnesium to barium.

_____

_____

_____

_____

[2 marks]

5.3   Give the balanced symbol equation for the reaction which occurs between solid barium and water. Include state symbols.

_____

[1 mark]

A student is conducting test tube reactions to deduce the identity of an aqueous solution. The student adds acidified barium chloride solution to a sample of the unknown solution to test for the presence of sulfate ions.

5.4   State the observation that would be made if sulfate ions were present in the solution.

_____

[1 mark]

The barium chloride solution is acidified with hydrochloric acid (HCl).

5.5   Explain why the barium chloride solution is acidified.

_____

_____

[1 mark]

5.6   Explain why sulfuric acid cannot be used to acidify the barium chloride solution.

_____

_____

[1 mark]

Turn over for the next question

Turn over ▶

6 This question is about Group 7 elements and their compounds.

When a sample of aqueous bromine is added to a solution of potassium iodide, a displacement reaction occurs.

$$Br_{2(aq)} + 2I^-_{(aq)} \rightarrow 2Br^-_{(aq)} + I_{2(aq)}$$

6.1 State the role of bromine in this reaction.

_____

[1 mark]

6.2 State the final colour of the solution formed in this reaction.

_____

[1 mark]

6.3 When chlorine gas is dissolved in a cold, dilute solution of sodium hydroxide, a redox reaction occurs in which **one** species is both oxidised and reduced.

$$Cl_{2(g)} + 2NaOH_{(aq)} \rightarrow NaCl_{(aq)} + NaClO_{(aq)} + H_2O_{(l)}$$

Identify the species that is both oxidised and reduced in the reaction above.

_____

[1 mark]

6.4 Hydrogen halides can be produced from the reaction between a sodium halide and concentrated sulfuric acid, as shown in the equation below.

$$NaX + H_2SO_4 \rightarrow NaHSO_4 + HX$$

A scientist measures the initial pH of some concentrated sulfuric acid, then adds 1 mol of solid NaCl to an excess of the acid. HCl gas is given off. After the reaction, the scientist measures the pH of the solution again and finds that it has increased. The scientist then repeats their experiment using 1 mol of solid NaBr.

Suggest why the pH of the solution produced from the reaction with NaBr is likely to be higher than the pH of the solution produced from the reaction with NaCl.

_____

_____

_____

_____

_____

_____

_____

_____

_____

_____

[4 marks]

6.5 A lab technician finds an unlabelled bottle containing a colourless solution.
The technician suspects that the solution is ammonium iodide.

Describe how the technician could use chemical tests to prove that the solution is ammonium iodide.

Leave blank

_____

_____

_____

_____

_____

_____

_____

_____

_____

_____

_____

_____

_____

_____

_____

_____

_____

_____

_____

_____

[6 marks]

Turn over for the next question

Turn over ▶

7 This question is about reactions in gaseous equilibria.

7.1 Ethanol is produced industrially in a gaseous equilibrium reaction where ethene is added to steam:

$$C_2H_{4(g)} + H_2O_{(g)} \rightleftharpoons C_2H_5OH_{(g)} \qquad \Delta H = -46 \text{ kJ mol}^{-1}$$

Use Le Chatelier's Principle to predict the effect on the equilibrium yield of ethanol if the overall pressure of this reaction is decreased. Explain your answer.

_____

_____

_____

_____

[3 marks]

Nitrogen and hydrogen gas can be combined to produce ammonia:

$$N_{2(g)} + 3H_{2(g)} \rightleftharpoons 2NH_{3(g)} \qquad \Delta H = -92 \text{ kJ mol}^{-1}$$

7.2 In an experiment, one mole of nitrogen gas and three moles of hydrogen gas were mixed in a sealed flask in the presence of an iron catalyst.
The mixture was left to reach equilibrium at a temperature of 500 K.
Analysis of the equilibrium mixture showed that 1.4 moles of ammonia had been produced.
The total pressure of the equilibrium mixture was 150 kPa.

Calculate the value of the equilibrium constant, K_p, at 500 K and state its units.

Value of K_p _____ Units _____

[7 marks]

7.3 State how a decrease in temperature will affect the value of K_p for this reaction. Justify your answer.

_____

_____

_____

_____

[3 marks]

Turn over for the next question

Turn over ▶

Table 3 shows some electrode potential data. Platinum electrodes were used in the half-cells where inert electrodes were required.

Table 3

Half-cell	Electrode potential, $E^{\ominus}$ (V)
$Mg^{2+}_{(aq)} + 2e^- \rightleftharpoons Mg_{(s)}$	−2.37
$V^{2+}_{(aq)} + 2e^- \rightleftharpoons V_{(s)}$	−1.18
$Zn^{2+}_{(aq)} + 2e^- \rightleftharpoons Zn_{(s)}$	−0.76
$V^{3+}_{(aq)} + e^- \rightleftharpoons V^{2+}_{(aq)}$	−0.26
$Cu^{2+}_{(aq)} + 2e^- \rightleftharpoons Cu_{(s)}$	+0.34
$VO^{2+}_{(aq)} + 2H^+_{(aq)} + e^- \rightleftharpoons V^{3+}_{(aq)} + H_2O_{(l)}$	+0.34
$Fe^{3+}_{(aq)} + e^- \rightleftharpoons Fe^{2+}_{(aq)}$	+0.77
$VO_2^+{}_{(aq)} + 2H^+_{(aq)} + e^- \rightleftharpoons VO^{2+}_{(aq)} + H_2O_{(l)}$	+1.00

8.1 Standard electrode potentials are measured under three standard conditions. State these standard conditions.

_____

_____

_____

[1 mark]

8.2 An electrochemical cell is made by connecting a Zn^{2+}/Zn half-cell to a Fe^{3+}/Fe^{2+} half-cell. Use the data shown in **Table 3** to give the conventional cell representation for the feasible reaction that occurs in this cell.

_____

[2 marks]

8.3 When excess granulated zinc is added to a yellow solution of ammonium vanadate(V) in acidic conditions, a series of redox reactions takes place.

Using data from **Table 3**, deduce the oxidation state of vanadium in the final vanadium species produced in these reactions. State the final colour of the solution. You should explain each stage of your deductions.

Oxidation state _____

Colour of solution _____

Explanation _____

_____

_____

_____

_____

_____

[5 marks]

8.4 Alkaline hydrogen-oxygen fuel cells can be used to power vehicles such as buses or cars. Suggest an advantage of using this type of fuel cell instead of fossil fuels to power these vehicles.

_____

_____

[1 mark]

Turn over for the next question

Turn over ▶

9 A student is conducting an experiment to investigate how the pH of a solution changes during an acid-base reaction. They slowly add 0.100 mol dm^{-3} sodium hydroxide solution, NaOH, to 25.0 cm^3 of 0.100 mol dm^{-3} ethanoic acid, CH$_3$COOH.

Leave blank

9.1 Calculate the pH of the 0.100 mol dm^{-3} solution of NaOH at 298K.
The ionic product of water, K_w = 1.0 x 10^{-14} mol^2 dm^{-6} at 298K.

pH _____

[2 marks]

9.2 Suggest a suitable piece of apparatus that could be used to monitor the pH during this reaction.

_____

[1 mark]

The results of the experiment are used to plot the pH curve shown in **Figure 1**.

Figure 1

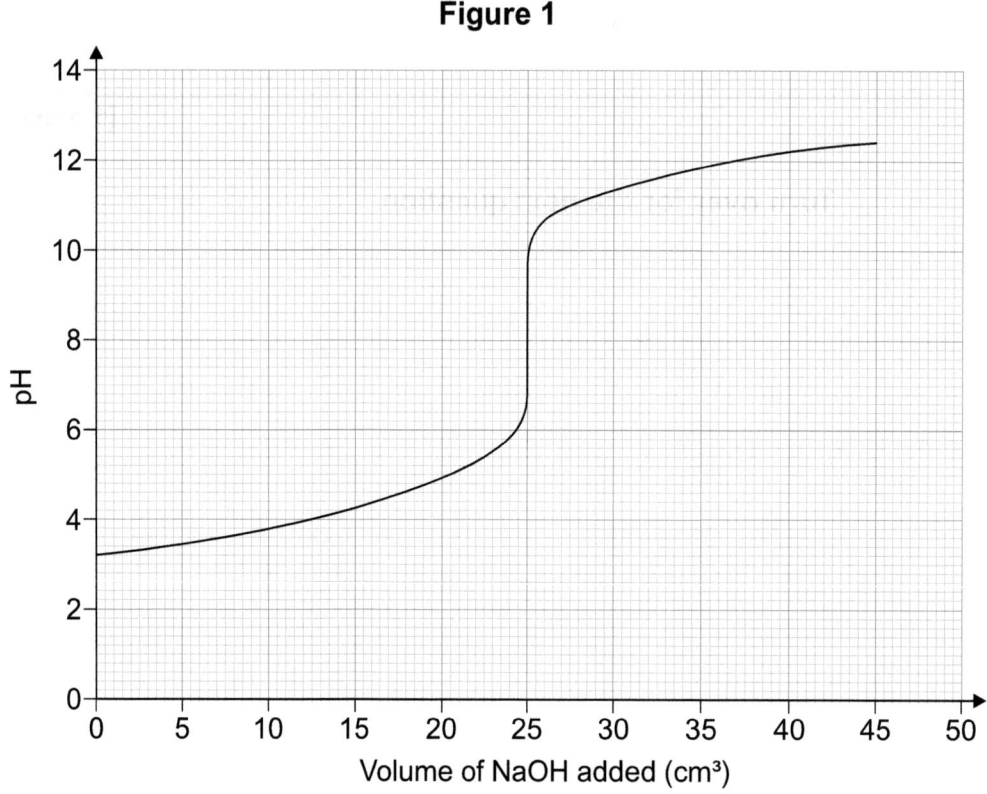

9.3 Ethanoic acid is a weak acid.
Give the expression for the acid dissociation constant, K_a, for ethanoic acid.

[1 mark]

9.4 Show that the pH of the solution will be equal to the pK_a when half of the acid has been neutralised.

_____

_____

_____

_____

_____

_____

_____

[2 marks]

9.5 Use **Figure 1** to deduce the volume of sodium hydroxide solution needed to neutralise half of the acid.

Volume _____ cm³

[1 mark]

Question 9 continues on the next page

Turn over ▶

9.6 Methanoic acid, HCOOH, is also a weak acid.
The value of K_a for HCOOH is 1.78×10^{-4} mol dm^{-3} at 298 K.
A buffer solution was prepared by adding 0.025 moles of HCOOH to 0.0125 moles of HCOONa in a total volume, V, of water.

Calculate the pH of the buffer solution formed at 298 K.
Give your answer to two decimal places.

pH _____

[4 marks]

10 This question is about thermochemical changes.

10.1 Magnesium bromide, $MgBr_2$, is an ionic compound which dissolves in water to produce aqueous magnesium ions and aqueous bromide ions.
Table 4 shows some enthalpy change values associated with this process.

Table 4

	Enthalpy Change (kJ mol^{-1})
$MgBr_{2(s)} \rightarrow Mg^{2+}_{(aq)} + 2Br^-_{(aq)}$	−154
$Mg^{2+}_{(g)} \rightarrow Mg^{2+}_{(aq)}$	−1920
$Br^-_{(g)} \rightarrow Br^-_{(aq)}$	−337

Use the data shown in **Table 4** to calculate the enthalpy of lattice formation for $MgBr_2$.

Enthalpy of lattice formation _____ kJ mol^{-1}
[2 marks]

Question 10 continues on the next page

Turn over ▶

10.2 The enthalpy of lattice formation for magnesium iodide, MgI_2, obtained by experiment is -2327 kJ mol^{-1}.
The value obtained by calculation using a perfect ionic model is -1944 kJ mol^{-1}.

Explain what these values suggest about the bonding in MgI_2.

_____

_____

_____

_____

[2 marks]

10.3 Sodium hydrogen carbonate undergoes a thermal decomposition reaction to produce sodium carbonate, water vapour and carbon dioxide gas as shown in the equation.

$$2NaHCO_{3(s)} \rightarrow Na_2CO_{3(s)} + H_2O_{(g)} + CO_{2(g)} \qquad \Delta H = +130 \text{ kJ mol}^{-1}$$

The entropy change for the reaction, $\Delta S = +335$ J K^{-1} mol^{-1}.

Calculate the minimum temperature in °C at which this reaction is feasible.
Give your answer to three significant figures.

Minimum temperature _____ °C

[3 marks]

11 This question is about transition metals and metal ions in aqueous solution.

Leave blank

11.1 Give the full electron configurations of a Cr atom and a Cr^{3+} ion.

Cr _____

Cr^{3+} _____

[2 marks]

11.2 The concentration of iron(II) ions in a solution can be determined by titration with acidified potassium manganate(VII) solution.

Deduce the overall ionic equation for the reaction of the iron(II) ions with manganate(VII) ions.

_____

[1 mark]

11.3 Write the balanced equation for the reaction that occurs when an excess of aqueous ammonia is added to a solution of aqueous iron(III) ions.
You must include state symbols in your equation.

_____

[2 marks]

11.4 Give the formula of the complex ion and the colour of the solution it forms when an excess of aqueous ammonia is added to a solution of aqueous copper(II) ions.

Formula _____

Colour _____

[2 marks]

Question 11 continues on the next page

Turn over ▶

11.5 Solution **X** contains an unidentified metal cation which may be Al^{3+} or Fe^{2+}.
A lab technician is asked to conduct test tube reactions to confirm the identity of the cation present in the solution.
A small sample of solution **X** is added to two separate test tubes.
The technician adds drops of sodium hydroxide solution continually to the first test tube, until it is in excess.
The technician then adds sodium carbonate solution to the second test tube.
The technician records any observations made during the reactions.

Explain how the technician could use the observations made during both of these reactions to deduce the identity of the cation present in solution **X**.

_____

_____

_____

_____

_____

_____

_____

_____

_____

_____

[4 marks]

END OF QUESTIONS

BLANK PAGE

CGP Practice Exam Paper
A-Level Chemistry

AQA A-Level Chemistry

Practice Paper 2 — Organic and Physical Chemistry

Time allowed: 2 hours

Centre name					
Centre number					
Candidate number					

Surname
Other names
Candidate signature

In addition to this paper you should have:
- A Data Booklet (including a Periodic Table)
- A ruler
- A calculator

Instructions to candidates
- Use black ink or ball-point pen.
- Write your name and other details in the spaces provided above.
- Answer all questions.
- Answer the questions in the spaces provided.
- Show clearly how you worked out your answers.
- Cross out any work you do not want to be marked.

Information for candidates
- There are 105 marks available for this paper.
- The marks available are given in brackets at the end of each question.

For examiner's use			
Q	Mark	Q	Mark
1		7	
2		8	
3		9	
4		10	
5		11	
6			
Total			

Answer ALL the questions.

Write your answers in the spaces provided.

1   This question is about alcohols.

Ethanol is a biofuel that can be produced by the fermentation of glucose from plants. Ethanol produced by this method is commonly considered to be a carbon-neutral fuel.

1.1   Using equations to support your answer, explain why ethanol produced by the fermentation of glucose can be considered a carbon-neutral fuel.

_____

_____

_____

_____

_____

_____

_____

_____

_____

[4 marks]

1.2   Give **one** reason why ethanol produced by the fermentation of glucose may **not** be considered a carbon-neutral fuel.

_____

_____

_____

[1 mark]

Pentan-1-ol is an alcohol commonly used as a solvent for coating CDs and DVDs.

1.3 Name a process that could be used to separate pentan-1-ol from a mixture of alcohols.

_____

[1 mark]

1.4 Briefly describe a chemical process that could be used to distinguish pentan-1-ol from pentan-2-ol.

_____

_____

_____

_____

_____

_____

_____

_____

_____

[3 marks]

Question 1 continues on the next page

Turn over ▶

Methanol is produced industrially from carbon monoxide and hydrogen.
The equation for this reaction is shown below:

$$CO_{(g)} + 2H_{2(g)} \rightleftharpoons CH_3OH_{(g)} \qquad \Delta H = -90 \text{ kJ mol}^{-1}$$

1.5 Write the expression for the equilibrium constant, K_c, for this reaction and give its units.

units _____

[2 marks]

1.6 The industrial production of methanol is carried out at a temperature of 250 °C and a pressure of 5000–10 000 kPa over a copper catalyst.

Explain this choice of reaction conditions in terms of their effect on reaction yield, reaction rate and cost.

_____

_____

_____

_____

_____

_____

_____

_____

_____

_____

_____

_____

_____

_____

_____

_____

_____

Leave blank

[6 marks]

Turn over for the next question

Turn over ▶

2   Nitric oxide, NO, reacts with oxygen, O_2, to produce nitrogen dioxide, NO_2.
The rate equation for this reaction is as follows:

$$Rate = k[NO]^2[O_2]$$

At a certain temperature, 5.0 mol dm^{-3} O_2 reacts with 0.75 mol dm^{-3} NO.
Figure 1 shows how the concentration of NO_2 changes with time during this reaction.

Figure 1

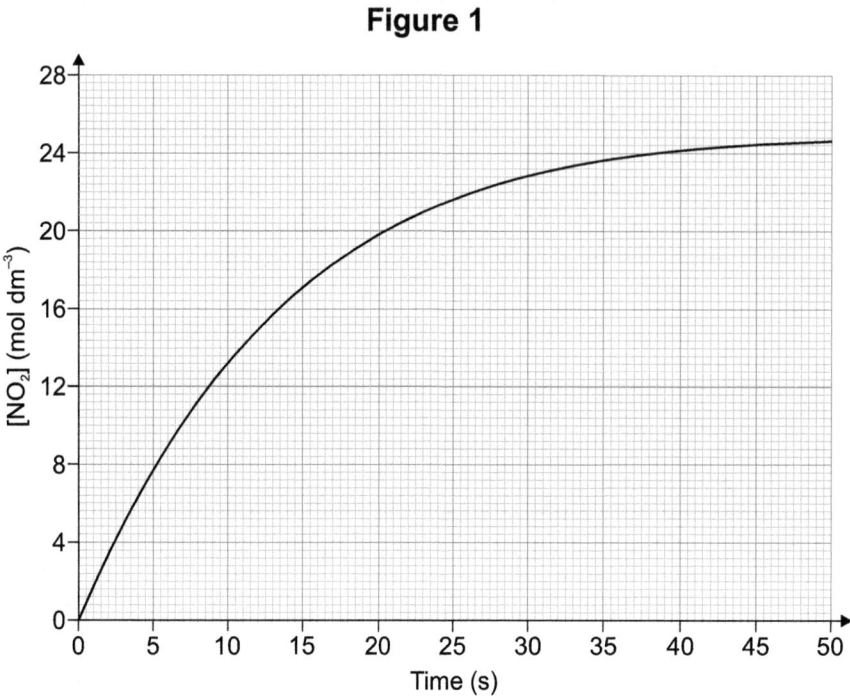

2.1   Using **Figure 1** and the information above, calculate the value of the rate constant, *k*, for this reaction and give its units.

rate constant, *k* _____ units _____

[5 marks]

2.2   The experiment was repeated using double the initial concentration of oxygen.
State the effect (if any) that this would have on the value of the rate constant.

_____

[1 mark]

Nitric oxide also reacts with hydrogen to produce nitrogen and water.
The results of a series of experiments investigating this reaction are shown in **Table 1**.
The temperature was kept constant for all of the experiments.

Table 1

Experiment	[NO] (mol dm^{-3})	[H$_2$] (mol dm^{-3})	Initial Rate (mol dm^{-3} s^{-1})
1	1.6×10^{-3}	8.0×10^{-4}	1.2×10^{-3}
2	1.6×10^{-3}	2.0×10^{-3}	3.0×10^{-3}
3	3.2×10^{-3}	1.6×10^{-3}	9.6×10^{-3}
4	4.8×10^{-3}	1.2×10^{-3}	

2.3　Determine the rate equation for this reaction.

rate equation _____

[2 marks]

2.4　Determine the initial rate for experiment **4**.

initial rate _____ mol dm^{-3} s^{-1}

[1 mark]

Turn over for the next question

Turn over ▶

3 Compound **A** is a very reactive molecule that can be used in organic synthesis. The structure of compound **A** and some of its reaction pathways are shown in **Figure 2**.

Figure 2

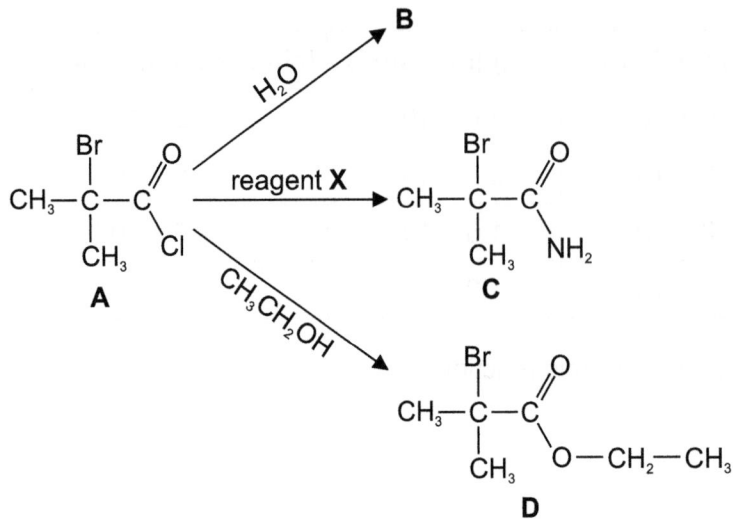

3.1 Give the IUPAC name for compound **A**.

_____

[1 mark]

3.2 Explain why compound **A** is very reactive.

_____

_____

_____

_____

[2 marks]

3.3 Name and outline the mechanism for the reaction of compound **A** with water to produce compound **B**.

Name _____

Mechanism

[5 marks]

3.4 Identify reagent **X**, which is required to convert compound **A** into compound **C**.

_____

[1 mark]

3.5 The reaction of compound **A** with ethanol produces compound **D**. How many peaks would you expect to observe in the ^{1}H NMR spectrum of compound **D**?

_____

[1 mark]

3.6 Draw an isomer of compound **D** that would react with sodium carbonate solution to produce carbon dioxide.

[1 mark]

Turn over for the next question

Turn over ▶

4 Ammonium chloride decomposes when heated to produce ammonia and hydrogen chloride. The equation for this reaction is shown below:

$$NH_4Cl_{(s)} \rightarrow NH_{3(g)} + HCl_{(g)} \qquad \Delta H = +176 \text{ kJ mol}^{-1}$$

4.1 Describe the different types of bonding that are present in solid ammonium chloride.

_____

_____

_____

_____

_____

_____

[3 marks]

4.2 Calculate the volume of ammonia gas that will be produced when 1.0 g of NH_4Cl completely decomposes at 350 °C and 100 kPa. Give your answer in dm³.
The gas constant, R = 8.31 J K⁻¹ mol⁻¹.

volume _____ dm³

[4 marks]

If ammonia is released into the atmosphere, it can catalyse the hydrolysis of sulfur trioxide. This produces sulfuric acid, which can contribute to the formation of acid rain.

4.3 Draw the shape of a sulfur trioxide molecule.
Show the approximate value of the bond angle.

[2 marks]

4.4 Sulfur trioxide can be produced from the following reaction between sulfur dioxide and oxygen in the atmosphere:

$$2SO_{2(g)} + O_{2(g)} \rightarrow 2SO_{3(g)}$$

Table 2

Bond	Mean bond enthalpy (kJ mol^{-1})
S=O	523
O–O	146
O=O	498

Using the mean bond enthalpies provided in **Table 2**, calculate the enthalpy change for the above reaction.

ΔH _____ kJ mol^{-1}
[2 marks]

Turn over for the next question

Turn over ▶

5 **Figure 3** shows the structure of four organic compounds.

Figure 3

E

F

G

H

5.1 Which of these compounds, if any, displays E/Z isomerism?

_____

[1 mark]

5.2 Describe a test-tube reaction that could be used to distinguish between compounds **E** and **F**.

_____

_____

_____

_____

_____

[2 marks]

5.3 Using the data booklet to help you, describe **two** differences that would be observed between the infrared spectra of compounds **E** and **H**.

_____

_____

_____

_____

_____

[2 marks]

The reaction of compound **E** with hydrogen bromide produces a mixture of compounds **F** and **G**.

5.4 Briefly explain which compound would be the major product of this reaction.

_____

_____

_____

[1 mark]

5.5 Explain why the mixture produced by this reaction contains three isomers, and the effect that this mixture would have on plane polarised light.

_____

_____

_____

_____

_____

_____

_____

_____

_____

_____

_____

_____

_____

_____

[5 marks]

Turn over for the next question

Turn over ▶

6 Aspartame is an artificial sweetener commonly used as a sugar substitute. It is the methyl ester of a dipeptide. The structure of aspartame and the dipeptide that it is based on are shown in **Figure 4**.

Figure 4

Aspartame   **Dipeptide**

6.1 Draw the structures of the **two** amino acids that would produce the dipeptide shown in **Figure 4**.

[2 marks]

6.2 Methyl esters are usually produced from carboxylic acids via esterification reactions. Give the type of catalyst typically used in esterification reactions.

_____

[1 mark]

6.3 Explain why performing the esterification of the dipeptide shown in **Figure 4** is **not** a good method for synthesising aspartame.

_____

_____

_____

[2 marks]

6.4 Cysteine is an amino acid that can be found in yoghurts, which typically have a pH of between 4 and 5. Draw the structure of cysteine when it is present in a yoghurt. Use the data booklet to help you.

[1 mark]

7 This question is about DNA.

A scientist is experimenting with a sample of DNA. In one experiment, the DNA is heated to temperature **X** to separate its polynucleotide strands.

7.1 State the type of bond that is broken during this process.

_____

[1 mark]

7.2 Explain why a higher value of **X** is required for a sample of DNA that contains a very high proportion of guanine/cytosine nucleotide pairs, compared with a sample of DNA containing a very high proportion of adenine/thymine nucleotide pairs.

_____

_____

_____

_____

_____

[2 marks]

7.3 Using the data booklet to help you, draw the structure of a free adenine nucleotide.

[2 marks]

7.4 Explain why the experiment would not work if the scientist added cisplatin to the DNA sample.

_____

_____

_____

_____

[2 marks]

Turn over for the next question

Turn over ▶

8 A group of students used calorimetry to measure the standard enthalpy of combustion of propanol. The equipment that they used is shown in **Figure 5**.

Figure 5

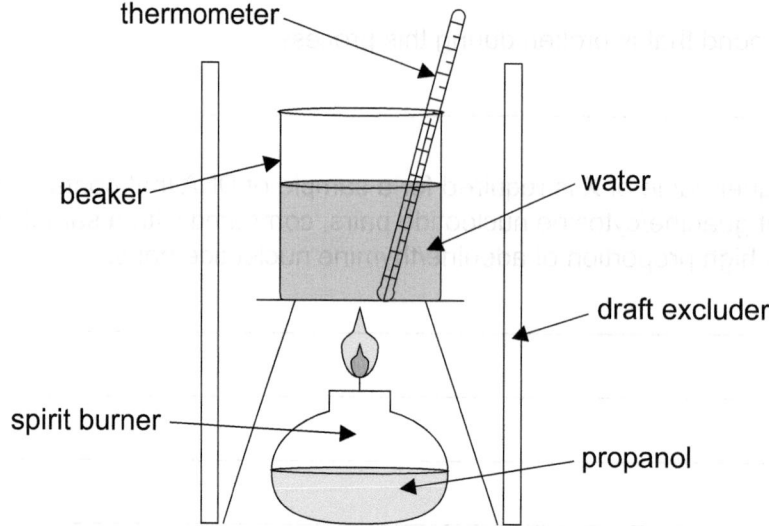

thermometer

beaker

water

draft excluder

spirit burner

propanol

The students burned 1.80 g of propanol completely in order to heat 200 g of water. The temperature of the water at the beginning of the experiment was 21.4 °C and the maximum temperature that they recorded was 76.6 °C.

8.1 Use the results from the students' experiment to calculate the standard enthalpy of combustion, $\Delta_c H$, of propanol in kJ mol^{-1}. Give your answer to three significant figures. The specific heat capacity of water is 4.18 J K^{-1} g^{-1}.

$\Delta_c H$ _____ kJ mol^{-1}
[3 marks]

8.2 **Table 3** shows the enthalpies of formation for propanol and the products of its combustion.

Table 3

Compound	$\Delta_f H$ (kJ mol⁻¹)
C_3H_7OH	−302.7
CO_2	−393.5
H_2O	−285.8

Use the data in **Table 3** to calculate the theoretical value of the enthalpy of combustion of propanol, in kJ mol⁻¹.

$\Delta_c H$ _____ kJ mol⁻¹

[3 marks]

8.3 Give **one** reason why the students' experimental value and the theoretical value calculated using Hess's law are not the same.

_____

_____

[1 mark]

8.4 Describe **one** modification that the students could have made to their experiment to give a more accurate value of $\Delta_c H$.

_____

_____

[1 mark]

8.5 Suggest why using 100 g of water in this experiment would not have given an accurate value of $\Delta_c H$.

_____

_____

_____

_____

[2 marks]

Turn over for the next question

Turn over ▶

9   **Figure 6** shows a pair of compounds, **J** and **K**. Compounds **J** and **K** are both monomers that can be used to make polymers.

Figure 6

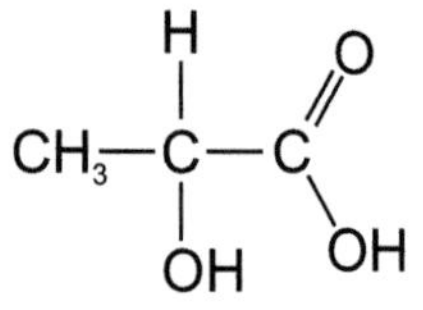

J 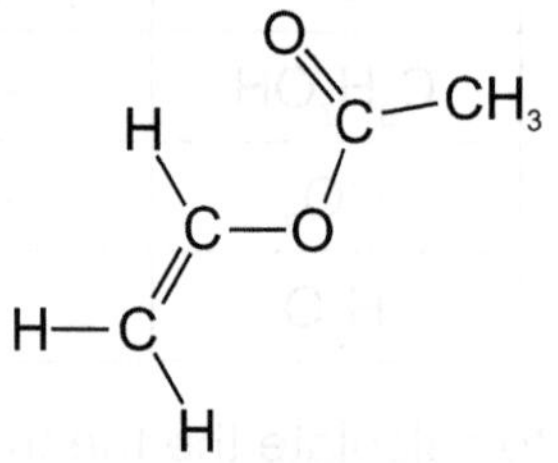   K

9.1   What type of polymer will compound **J** produce? Tick **one** box.

A   Polyamide   ☐

B   Polyalkene   ☐

C   Polyester   ☐

D   Polypeptide   ☐

[1 mark]

9.2   Draw the repeating unit of the polymer produced from compound **K**.

[1 mark]

9.3 The polymer produced from compound **K** can be used as a component in chewing gum. Give **one** disadvantage of using this polymer in chewing gum.

_____

_____

[1 mark]

9.4 The polymer produced from compound **K** can be converted into a new polymer, **L**, by reaction with water in the presence of an acid catalyst.
Draw the repeating unit of polymer **L** and identify the extra product of the reaction.

Extra product _____

[2 marks]

Turn over for the next question

Turn over ▶

10 Phenylethanamide is an important precursor in the synthesis of a number of pharmaceuticals. It can be synthesised in a three-step synthesis starting from benzene, as shown in **Figure 7**.

Figure 7

10.1 In Step **1** of the synthesis, benzene reacts with a mixture of nitric acid, HNO_3, and sulfuric acid, H_2SO_4.

Outline the mechanism for Step **1** of this synthesis, including equations to show the formation of the electrophile.

Formation of electrophile _____

_____

[4 marks]

10.2 Explain why benzene typically undergoes substitution reactions rather than addition reactions.

_____

_____

_____

[1 mark]

10.3 Outline the conditions required for Step **2** of this synthesis.

_____

_____

[2 marks]

Step **3** of this synthesis can be performed using **either** of the **two** reactions below:

Reaction **A**: $C_6H_5NH_2 + (CH_3CO)_2O \rightarrow C_6H_5NHCOCH_3 + CH_3COOH$
Reaction **B**: $C_6H_5NH_2 + CH_3COCl \rightarrow C_6H_5NHCOCH_3 + HCl$

10.4 Calculate the atom economy for reaction **A**.

Atom economy _____ %

[2 marks]

10.5 Suggest why reaction **A** is preferred industrially, even though reaction **B** has a greater atom economy.

_____

_____

[1 mark]

Turn over for the next question

Turn over ▶

11 An unknown organic compound that can cause allergic reactions in some users has been isolated from a sample of nail varnish. Using mass spectrometry, scientists determined that the *m/z* of the molecular ion of this compound is 86.0448.
The infrared spectrum (**Figure 8**) and ^{13}C NMR spectrum (**Figure 9**) of this compound are shown below.

Figure 8

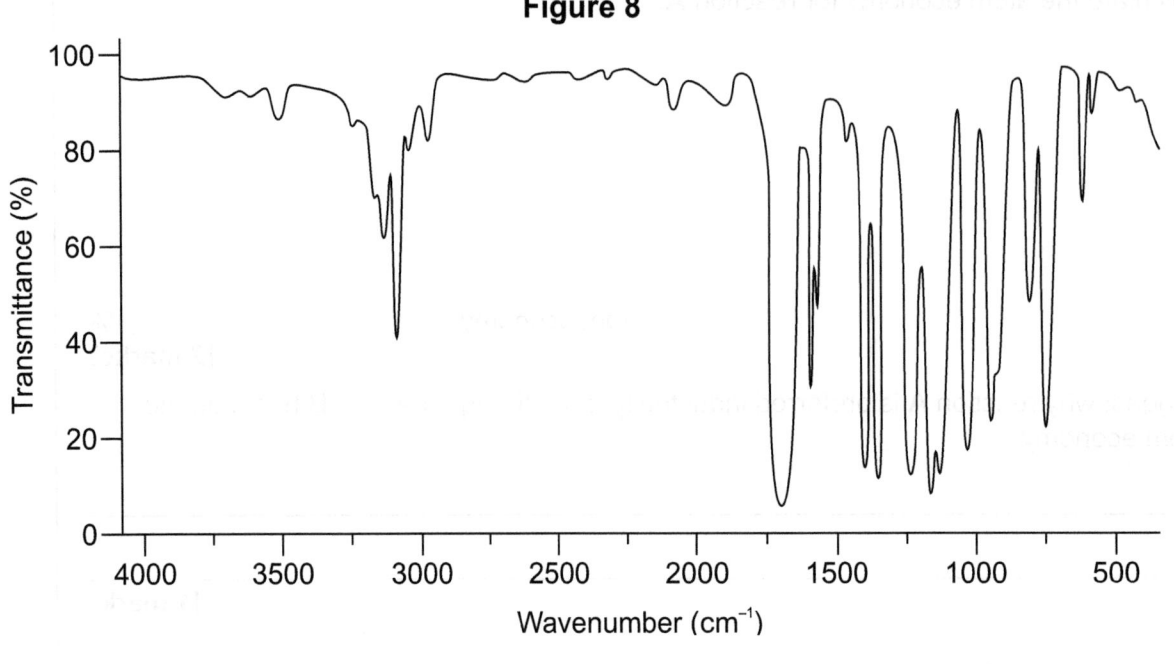

Figure 9

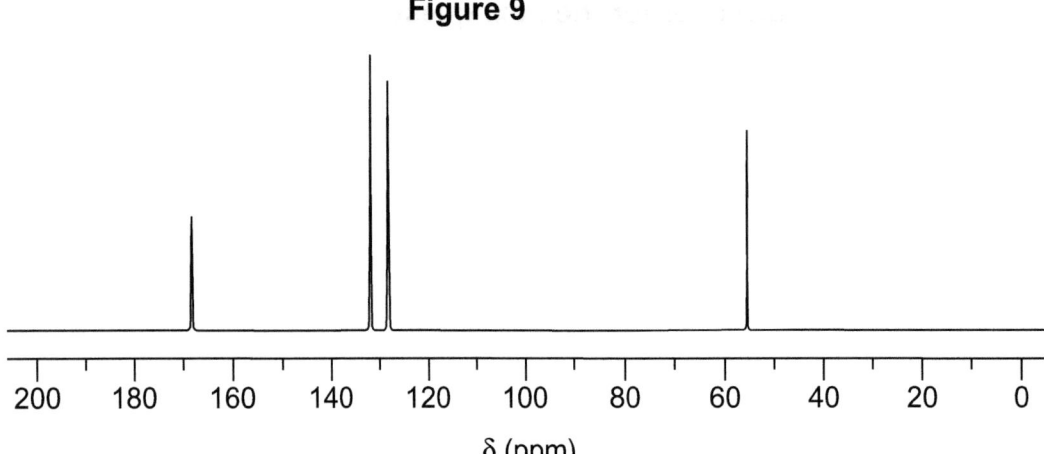

Using the information above, determine the **two** possible structures of the unknown organic compound. Fully explain how you deduced your answer.
The following precise atomic masses can be used:
^{1}H: 1.0078, ^{12}C: 12.0000, ^{14}N: 14.0064, ^{16}O: 15.9990.

Structures

Explanation _____

_____

_____

_____

_____

_____

_____

_____

_____

_____

_____

_____

_____

_____

_____

_____

_____

_____

_____

_____

_____

_____

_____

_____

_____

_____

_____

[8 marks]

END OF QUESTIONS

AQA A-Level Chemistry

Practice Paper 3

Time allowed: 2 hours

Centre name					
Centre number					
Candidate number					

Surname
Other names
Candidate signature

In addition to this paper you should have:
- A Data Booklet (including a Periodic Table)
- A ruler
- A calculator

Instructions to candidates
- Use black ink or ball-point pen.
- Write your name and other details in the spaces provided above.
- Answer all questions.
- Answer the questions in the spaces provided.
- Show clearly how you worked out your answers.
- Cross out any work you do not want to be marked.

Information for candidates
- There are 90 marks available for this paper.
- The marks available are given in brackets at the end of each question.
- There are two sections in this paper. It is recommended that you spend approximately 70 minutes on Section A and 50 minutes on Section B.

For examiner's use	
Q	Mark
Section A	
1	
2	
3	
4	
Section B	
5 to 34	
Total	

Section A

Answer ALL the questions in this section.

Write your answers in the spaces provided.

1 Esters are sweet smelling liquids often used in food flavourings and fragrances.
The structure of an ester with molecular formula $C_7H_{14}O_2$ is shown in **Figure 1**.

Figure 1

1.1 This ester exists as a pair of stereoisomers.
State how the structure of one stereoisomer is related to the other and circle the chiral carbon in **Figure 1**.

_____

_____

_____

[2 marks]

The ester in **Figure 1** can be prepared in the laboratory by heating an alcohol and a carboxylic acid under reflux in the presence of a strong acid catalyst.
A student researched this reaction and obtained a method to carry out in the laboratory.

1.2 Name the carboxylic acid used to produce this ester.

_____

[1 mark]

1.3 After refluxing, the method suggested transferring the product to a separating funnel and shaking with sodium carbonate.

State the purpose of adding sodium carbonate to the product and suggest a precaution that the student should take after shaking the reaction mixture with sodium carbonate in the separating funnel.

Purpose _____

_____

Precaution _____

_____

[2 marks]

After running off the lower aqueous layer, the method suggested transferring the organic layer to a small stoppered conical flask and shaking with granules of calcium chloride.

1.4 State the role of the calcium chloride.

_____

_____

[1 mark]

1.5 Suggest how the calcium chloride could be removed from the product.

_____

_____

[1 mark]

Question 1 continues on the next page

Turn over ▶

Figure 1 is shown again below to help you answer the following questions.

Figure 1

The student's method then recommended using distillation apparatus to obtain a pure sample of the ester. Data book values for the boiling points of some components of the reaction are given in **Table 1**.

Table 1

Component of mixture	Boiling point (°C)
Ester	133
Carboxylic acid	176
Alcohol	78

1.6 The student predicted that the ester would have a lower boiling point than the alcohol.

Give **one** possible reason why the student may have made this prediction and use your understanding of structure and bonding to suggest why they were incorrect.

_____

_____

_____

_____

_____

_____

_____

_____

_____

[3 marks]

1.7 Use the information in **Table 1** to briefly describe how distillation could be used to obtain a pure sample of the ester from the reaction mixture. Include in your answer how you could confirm that the sample of the ester is pure.

_____

_____

_____

_____

_____

_____

_____

_____

_____

[3 marks]

1.8 For the esterification reaction, the student used 8.50 g of an aqueous solution of the appropriate alcohol. The concentration of the solution was 78.5% alcohol by mass.

Use **Figure 1** to deduce the relative molecular mass of the alcohol and hence calculate the number of moles of alcohol present in the 8.50 g solution.
Give your answer to three significant figures.

Number of moles _____

[3 marks]

The general equation for an esterification reaction is:

carboxylic acid + alcohol → ester + water

1.9 Calculate the atom economy for the student's reaction.
Give your answer to three significant figures.

Atom economy _____ %

[2 marks]

Turn over for the next question

Turn over ▶

2 Propanone and iodine will react together in acidic solution.
During the reaction, the brown colour of the iodine fades as it is converted to colourless iodide ions. The rate of this reaction can be monitored using a colorimeter.
The overall equation for the reaction is given below:

$$I_{2(aq)} + CH_3COCH_{3(aq)} \rightarrow CH_3COCH_2I_{(aq)} + H^+_{(aq)} + I^-_{(aq)}$$

2.1 Suggest the type of reaction that occurs when propanone reacts with iodine in acidic solution.

_____

[1 mark]

A scientist used a colorimeter, an appropriate coloured filter and iodine solutions of known concentration to determine the absorbance at different concentrations of iodine. They started with a stock solution of 0.0040 mol dm⁻³ iodine, and used their results to plot a calibration curve. **Figure 2** shows the calibration curve.

Figure 2

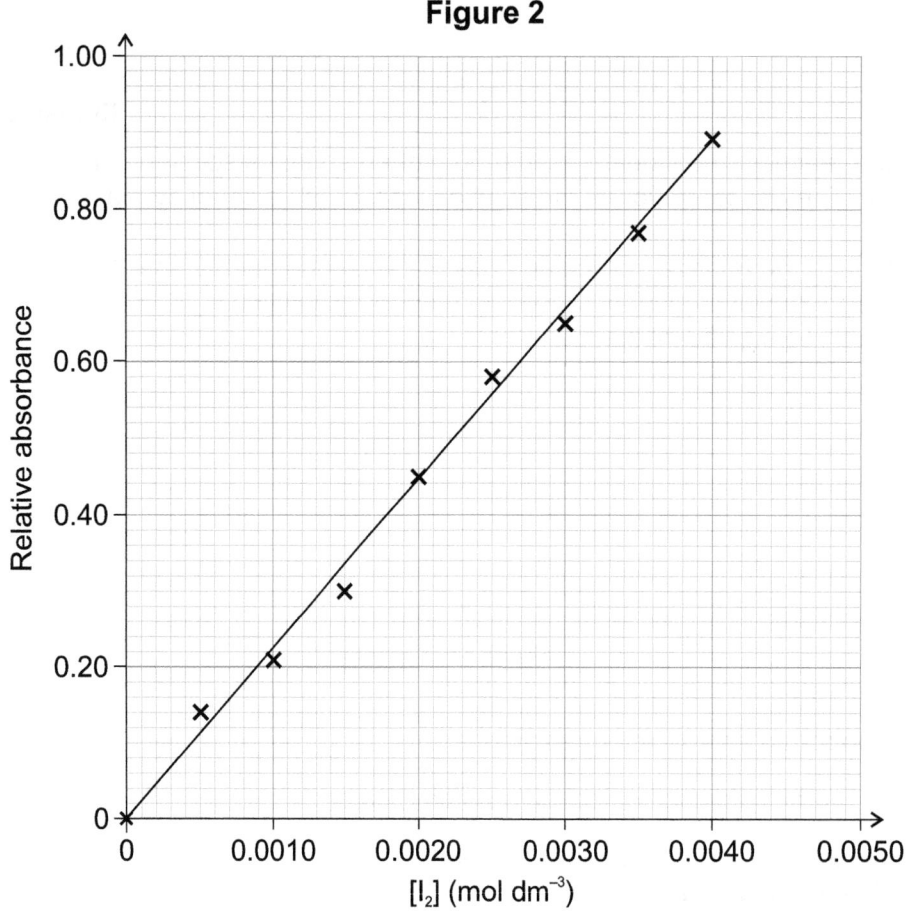

The scientist then added 20 cm³ of 0.50 mol dm⁻³ propanone to 10 cm³ of 1.0 mol dm⁻³ hydrochloric acid. They added this mixture to 30 cm³ of the stock iodine solution in a conical flask, and immediately started a timer. They removed small samples from the reaction mixture at regular intervals to be analysed in the colorimeter.

2.2 Explain why the scientist chose concentrations of propanone and hydrochloric acid that were much greater than that of the iodine solution.

_____

_____

[1 mark]

The results of this experiment are shown in **Table 2**:

Table 2

Time (s)	Relative Absorbance	[I_2] (mol dm^{-3})
0	0.89	
30	0.76	
60	0.60	
90	0.47	
120	0.34	
150	0.20	

2.3 Use the calibration curve in **Figure 2** to complete **Table 2**.
Plot a graph of iodine concentration against time on the grid below.
Draw a line of best fit on your graph.

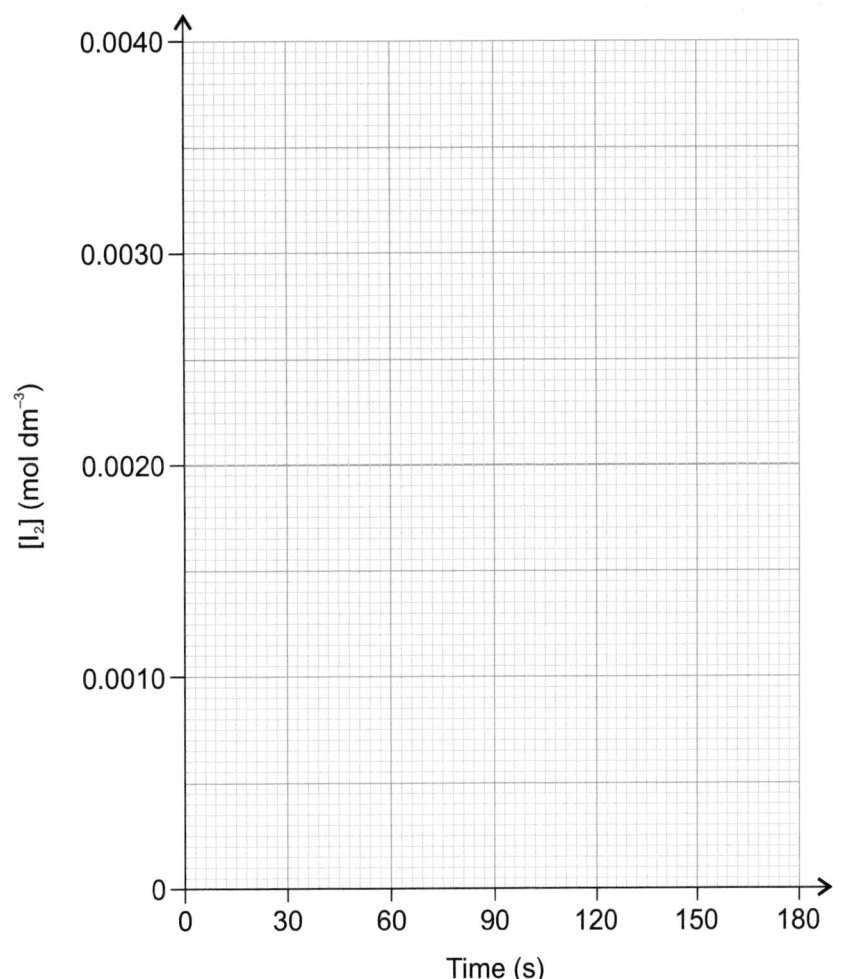

[3 marks]

2.4 Use your graph to deduce the order of reaction with respect to iodine.

_____

[1 mark]

Question 2 continues on the next page

Turn over ▶

2.5 The reaction between propanone and iodine proceeds via a multiple-step mechanism. The slowest step in the mechanism is the first step, shown below:

$$CH_3COCH_3 + H^+ \rightarrow CH_3C^+(OH)CH_3$$

Use the information given and your graph to deduce a rate equation for the overall reaction between propanone and iodine. Give the units for the rate constant, k.

Rate equation _____

Units of k _____

[2 marks]

2.6 The order of a reaction with respect to a particular reactant can also be determined using an initial rates method. The reaction between NO and Cl_2 is second order with respect to NO. The equation for this reaction is:

$$2NO_{(g)} + Cl_{2(g)} \rightarrow 2NOCl_{(g)}$$

Describe how you would use an initial rates method to determine that this reaction is second order with respect to NO. Assume that the concentrations of Cl_2 and NO can both be measured throughout the reaction.

_____

_____

_____

_____

_____

_____

_____

_____

_____

_____

_____

_____

_____

_____

_____

_____

_____

_____

_____

_____

_____

_____

_____

_____

_____

_____

_____

_____

_____

_____

_____

_____

_____

_____

_____

_____

_____

_____

_____

_____

_____

_____

[6 marks]

Turn over for the next question

Turn over ▶

3 Two reactions of a halogenoalkane are shown in **Figure 3**.

Leave blank

Figure 3

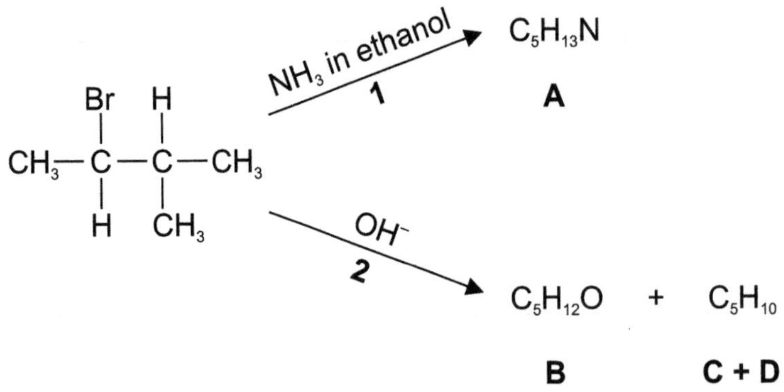

3.1 Suggest **one** reason why reaction **1** is not carried out under reflux.

[1 mark]

3.2 Draw the structure of an isomer of compound **A** that is a secondary amine and has three peaks in its ^{13}C NMR spectrum.

[1 mark]

3.3 Compound **A** is a weak base, but a stronger base than ammonia.
Explain this difference in base strength.

[2 marks]

3.4 Briefly explain why reaction **2** can produce **either** compound **B or** a mixture of compounds **C** and **D**.

[2 marks]

3.5 A student carried out reaction **2**. They obtained an infrared spectrum of their product, shown in **Figure 4**.

Figure 4

Describe how the spectrum in **Figure 4** would compare to an infrared spectrum of pure compound **B**.

_____

_____

_____

_____

_____

_____

_____

_____

[3 marks]

3.6 Describe how the student could confirm the presence of compound **B** in their product using a chemical test.

_____

_____

_____

_____

_____

[2 marks]

Turn over for the next question

Turn over ▶

4 Aluminium sulfate solution and aluminium chloride solution both contain the metal-aqua ion $[Al(H_2O)_6]^{3+}$. In solution, the metal-aqua ion undergoes a hydrolysis reaction, which can be simplified to:

$$[Al(H_2O)_6]^{3+}_{(aq)} \rightleftharpoons [Al(H_2O)_5(OH)]^{2+}_{(aq)} + H^+_{(aq)}$$

Write an expression for the acid dissociation constant, K_a, for $[Al(H_2O)_6]^{3+}$.

[1 mark]

4.2 The pK_a of $[Al(H_2O)_6]^{3+}$ is 4.85. The pH of an aluminium chloride solution was measured with a calibrated pH probe and found to be 2.81.

Calculate the concentration of the aluminium chloride solution.
Give your answer to three significant figures.

Concentration _____ mol dm^{-3}
[3 marks]

An excess of acidified silver nitrate solution was added to 20.0 cm^3 of the aluminium chloride solution. The precipitate that formed was collected on fluted filter paper, washed and dried before being weighed to calculate the percentage yield.

4.3 The percentage yield for the reaction was 32.0%. The equation for the reaction is:

$$AlCl_3 + 3AgNO_3 \rightarrow 3AgCl + Al(NO_3)_3$$

Use your answer to question **4.2** to calculate the mass of the dry silver chloride precipitate. If you did not obtain an answer to question **4.2**, assume that the concentration of the aluminium chloride solution is 0.191 mol dm^{-3}. (This is **not** the correct answer.)

Mass _____ g
[3 marks]

A-Level Chemistry / AQA / Paper 3   12   © CGP 2020 — copying more than 5% of this paper is not permitted

4.4 State the colour of the precipitate formed. Suggest a reason why it was collected on **fluted** filter paper.

Colour _____

Reason _____

_____

[2 marks]

4.5 Explain why the precipitate was washed **and** dried before being weighed.

_____

_____

[2 marks]

A student prepared a 250 cm^3 standard solution of 0.0150 mol dm^{-3} aluminium sulfate.

4.6 Calculate the mass of aluminium sulfate, $Al_2(SO_4)_3$, in this standard solution.

Mass _____ g

[2 marks]

The student used the following method to prepare the standard solution:
1. Weigh out the solid aluminium sulfate on a balance accurate to 3 decimal places.
2. Transfer the solid aluminium sulfate to a beaker, add distilled water and stir to ensure all the solid has dissolved.
3. Use a filter funnel to transfer the solution to a 250 cm^3 volumetric flask and add distilled water until the bottom of the meniscus is level with the graduation line of the flask.

4.7 Suggest **two** additional steps the student should include in the preparation of the standard solution to ensure the concentration is accurate.

1 _____

_____

2 _____

_____

[2 marks]

4.8 Sodium carbonate solution was added dropwise to a small portion of the standard aluminium sulfate solution in a test tube.

State what you would observe during this reaction and identify the products.

Observation(s) _____

Products _____

[2 marks]

Turn over for the next question　　　　　　　**Turn over ▶**

Section B

Answer ALL the questions in this section.

Tick one box per question.

---

5   Which of the following shows the molecules in the correct order from lowest to highest boiling point?

    A $HCl < HBr < HF$ ☐

    B $CH_3CHO < CH_3COOH < CH_3CH_2OH$ ☐

    C $CH_4 < C_3H_8 < C_2H_6$ ☐

    D $I_2 < Br_2 < Cl_2$ ☐

           [1 mark]

6   Which compound in **Table 3** has the most covalent character?

Table 3

Compound	Lattice enthalpy from Born Haber cycle (kJ mol⁻¹)	Theoretical lattice enthalpy (kJ mol⁻¹)	
A	673	670	☐
B	659	645	☐
C	756	732	☐
D	782	769	☐

           [1 mark]

7   Which of the following gives the correct catalyst for the stated reaction?

    A   A synthetic zeolite for thermal cracking. ☐

    B $ClO^•$ for the breakdown of ozone. ☐

    C Mn^{3+} ions for the reaction between aqueous MnO_4^- and $C_2O_4^{2-}$ ions. ☐

    D Fe^{2+} ions for the reaction between aqueous I^- and $S_2O_8^{2-}$ ions. ☐

           [1 mark]

8 Thin layer chromatography was used to analyse **two** samples, **V** and **W**, each known to contain a mixture of amino acids. The chromatogram produced is shown in **Figure 5**.

Figure 5

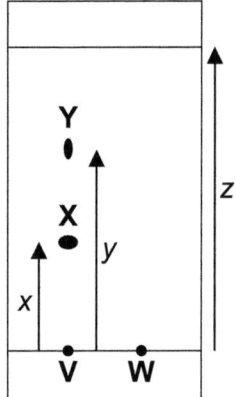

Which of the following is a correct statement?

A The plate should have been left in the solvent for longer to allow the solvent front to reach the top of the plate. ☐

B Sample **W** is more soluble in the solvent than sample **V**. ☐

C Amino acid **Y** has a lower retention in the stationary phase than amino acid **X**. ☐

D The R_f value for amino acid **X** can be calculated using the expression $\frac{z}{x}$. ☐

[1 mark]

9 Some zinc powder was reacted with 40.0 cm³ of a copper sulfate solution, and the increase in temperature was recorded as 16.2 °C.

What is the value, to three significant figures, of the heat energy, *q*, given out in this reaction?
You may assume that only the water changes in temperature and that the specific heat capacity of water is 4.18 J K⁻¹ g⁻¹.

A −2710 kJ ☐

B 2710 J ☐

C 2710 kJ ☐

D −2.71 J ☐

[1 mark]

Turn over for the next question

Turn over ▶

10   Which of the following molecules exists as a pair of stereoisomers?

   A   2-methylbut-2-ene   ☐

   B   2-fluorobut-1-ene   ☐

   C   2-methylbut-1-ene   ☐

   D   1-fluoro-2-methylbut-1-ene   ☐

   [1 mark]

11   When zinc metal is added to an acidic solution of vanadium(V) ions, a series of redox reactions takes place.

   Which statement about this process is correct?

   A   Zinc acts as a catalyst for the process.   ☐

   B   The final vanadium species produced contains V^{2+} ions.   ☐

   C   Zinc is reduced during the reaction.   ☐

   D   The initial colour of the vanadium(V) solution is green.   ☐

   [1 mark]

12   **Table 4** shows some electrode potentials.

Table 4

Electrode half-equation	$E^{\ominus}$ (V)
$Fe^{3+}_{(aq)} + e^- \rightleftharpoons Fe^{2+}_{(aq)}$	+0.77
$Zn^{2+}_{(aq)} + 2e^- \rightleftharpoons Zn_{(s)}$	−0.76
$Fe^{2+}_{(aq)} + 2e^- \rightleftharpoons Fe_{(s)}$	−0.44
$Cr_2O_7^{2-}_{(aq)} + 14H^+_{(aq)} + 6e^- \rightleftharpoons 2Cr^{3+}_{(aq)} + 7H_2O_{(l)}$	+1.33

Which statement is **not** correct?

   A   The conventional cell notation for the cell with an EMF of +0.56 V is: ☐

   $Pt \mid Fe^{2+}_{(aq)}, Fe^{3+}_{(aq)} \parallel Cr_2O_7^{2-}_{(aq)}, Cr^{3+}_{(aq)} \mid Pt$

   B   Zinc reacts with iron(II) ions to form iron.   ☐

   C   Zinc is a more powerful reducing agent than Cr^{3+}.   ☐

   D   The cell with an EMF of +0.32 V requires an inert electrode.   ☐

   [1 mark]

13 One mole of which species will react with 6.5 moles of oxygen to produce carbon dioxide and water?

A $CH_3CH_2CH_2CHO$ ☐

B $CH_3CH_2CH_2CH_2OH$ ☐

C C_4H_{10} ☐

D C_4H_8 ☐

[1 mark]

14 A student used a balance to find the mass of a sample of benzoic acid. The balance was accurate to three decimal places. They measured a value of 4.798 g for the empty weighing boat and 5.244 g for the weighing boat with the sample added.

If the uncertainty associated with the balance is 0.0005 g, what is the percentage uncertainty in the mass of the sample weighed out?

A   0.224% ☐

B   0.112% ☐

C   0.019% ☐

D   0.021% ☐

[1 mark]

15 Which row of **Table 5** shows the correct trends down Group 2?

Table 5

	Solubility of sulfates	Reactivity with water	Solubility of hydroxides	1st ionisation energy	
A	Increases	Decreases	Decreases	Increases	☐
B	Decreases	Increases	Increases	Decreases	☐
C	Decreases	Increases	Decreases	Increases	☐
D	Increases	Decreases	Increases	Decreases	☐

[1 mark]

Turn over for the next question

Turn over ▶

Which of the following is **not** a correct statement about the molecule shown in **Figure 6**?

Figure 6

$$CH_3-CH_2-N\begin{array}{c} H \\ CH_3 \end{array}$$

A   It can form a co-ordinate bond with a hydrogen ion.   ☐

B   It will react with bromomethane to form $CH_3CH_2N(CH_3)_2$.   ☐

C   It is a secondary amine.   ☐

D   It can be formed by the reduction of $CH_3CH_2CH_2NH_2$.   ☐

[1 mark]

17   What is the pH of a 0.380 mol dm^{-3} solution of barium hydroxide at 298 K?
($K_w = 1.00 \times 10^{-14}$ mol^2 dm^{-6})

A   13.6   ☐

B   13.9   ☐

C   14.0   ☐

D   14.1   ☐

[1 mark]

18   A bottle of household bleach contains 4.20 g of sodium chlorate(I), NaOCl, per 100 g.
100 g of the bleach occupies a volume of 90.1 cm^3.

What is the concentration of ClO$^-$ ions in the bleach?

A   1.25 mol dm^{-3} ☐

B   0.56 mol dm^{-3} ☐

C   0.63 mol dm^{-3} ☐

D   0.51 mol dm^{-3} ☐

[1 mark]

19 Which of the aromatic compounds shown below has the fewest peaks in its ^{13}C NMR spectrum?

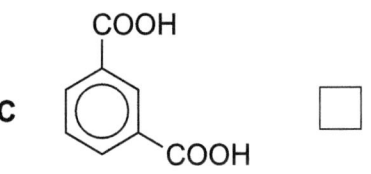

A ☐

B ☐

C ☐

D ☐

[1 mark]

20 Magnesium (A_r = 24.3) naturally occurs as three isotopes, ^{24}Mg, ^{25}Mg and ^{26}Mg. A sample of naturally-occurring magnesium is ionised by electron impact and travels through a time of flight mass spectrometer. A mass spectrum is recorded.

Which of the following statements about this experiment is correct?

A The $^{26}Mg^+$ ions drift faster than the $^{24}Mg^+$ ions. ☐

B The $^{24}Mg^+$ ions produce the tallest peak on the mass spectrum. ☐

C The $^{24}Mg^+$ ions spend the longest time in the drift region of the flight tube. ☐

D During ionisation, a high voltage is applied to the sample. ☐

[1 mark]

21 Which of these elements has the lowest first ionisation energy?

A Silicon ☐

B Aluminium ☐

C Sulfur ☐

D Phosphorus ☐

[1 mark]

Turn over for the next question

Turn over ▶

Given the information in **Table 6**, at what temperature (in °C) does the following process become feasible?

$$NH_{3(l)} \rightarrow NH_{3(g)}$$

Table 6

$\Delta H(NH_{3(l)} \rightarrow NH_{3(g)})$	$S(NH_{3(l)})$	$S(NH_{3(g)})$
23.4 kJ mol^{-1}	95.6 J K^{-1} mol^{-1}	192 J K^{-1} mol^{-1}

A −30.3 °C ☐

B 243 °C ☐

C 30.3 °C ☐

D −273 °C ☐

[1 mark]

23 Glucose can be converted to poly(ethene) in a three step process.
Which row in **Table 7** shows the correct reaction sequence for the process?

Table 7

	Reaction 1	**Reaction 2**	**Reaction 3**	
A	Fermentation	Elimination	Addition polymerisation	☐
B	Fermentation	Hydration	Addition polymerisation	☐
C	Fermentation	Elimination	Condensation polymerisation	☐
D	Fermentation	Hydration	Condensation polymerisation	☐

[1 mark]

24 0.897 g of a vaporised organic compound was found to occupy a volume of 373 cm^3 at a temperature of 403 K and a pressure of 101 kPa.

What is the relative molecular mass of the organic compound?
The gas constant, R = 8.31 J mol^{-1} K^{-1}.

A 134 ☐

B 25.7 ☐

C 78.0 ☐

D 79.7 ☐

[1 mark]

25 The rate constant, k, of a reaction is related to the activation energy, E_a, by the equation:

$$\ln(k) = -\frac{E_a}{RT} + \ln(A)$$

Where T is the temperature in Kelvin, A is the Arrhenius constant and R is the gas constant, $8.31\ J\ mol^{-1}\ K^{-1}$.

What is the correct sequence of steps for using this equation to calculate the activation energy of a reaction in $kJ\ mol^{-1}$?

A Plot a graph of $\ln(k)$ on the *y*-axis against $\frac{1}{T}$ on the *x*-axis.
Determine the gradient, multiply by –8.31 and divide by 1000. ☐

B Plot a graph of $\ln(k)$ on the *y*-axis against $\frac{1}{T}$ on the *x*-axis.
Determine the gradient and divide by 1000. ☐

C Plot a graph of $\frac{1}{T}$ on the *y*-axis against $\ln(k)$ on the *x*-axis.
Determine the gradient, multiply by 8.31 and divide by 1000. ☐

D Plot a graph of $\frac{1}{T}$ on the *y*-axis against $\ln(k)$ on the *x*-axis.
Determine the gradient and multiply by –8.31. ☐

[1 mark]

26 25.0 cm³ of an iron(II) solution was titrated against an acidified solution of potassium manganate(VII). The concentration of potassium manganate(VII) solution was 0.020 mol dm⁻³. The results are shown in **Table 8**.

Table 8

Titration	1	2	3	4
Initial burette reading (cm³)	0.00	0.40	0.10	0.50
Final burette reading (cm³)	16.20	15.80	16.00	15.80

The equation for the reaction of iron(II) ions with manganate(VII) ions is shown below:

$$5Fe^{2+}_{(aq)} + MnO_4^{-}{}_{(aq)} + 8H^{+}_{(aq)} \rightarrow 5Fe^{3+}_{(aq)} + Mn^{2+}_{(aq)} + 4H_2O_{(l)}$$

What is the mass of iron(II) ions present in the solution?

A 0.0857 g ☐

B 0.0882 g ☐

C 0.0176 g ☐

D 0.0171 g ☐

[1 mark]

Turn over for the next question

Turn over ▶

27   A mixture of three gases, **X**, **Y** and **Z**, contains 0.350 moles of gas **X**.
The total pressure of the system is 70.0 kPa and the partial pressure of gas **X** is 16.2 kPa.

What is the total number of moles of gas present in the system?

- **A**   1.42 ☐
- **B**   1.51 ☐
- **C**   1.89 ☐
- **D**   2.04 ☐

[1 mark]

28   The molecule vanillin is shown in **Figure 7**.

Figure 7

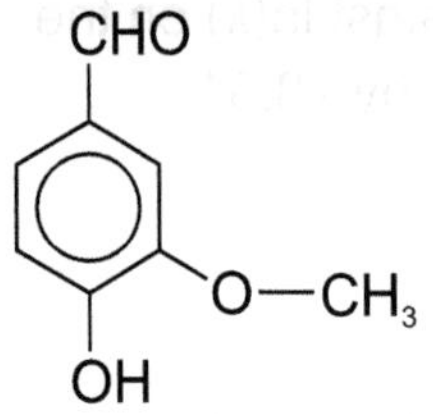

Which of the following is **not** a correct statement about vanillin?

- **A**   It turns acidified potassium dichromate from orange to green. ☐
- **B**   It forms a silver mirror when treated with Tollens' reagent. ☐
- **C**   It exists as a pair of enantiomers. ☐
- **D**   Its molecular formula is $C_8H_8O_3$. ☐

[1 mark]

29   Which of the following is a correct statement about copper ions in solution?

- **A** Cu^{2+} forms a square planar complex with chloride ions. ☐
- **B**   The complex formed by copper(II) and $C_2O_4^{2-}$ ions has a coordination number of 3. ☐
- **C**   Aqueous copper ions undergo a partial substitution reaction with ammonia solution. ☐
- **D**   The complex formed by Cu^{2+} and $EDTA^{4-}$ ions is less stable than the metal aqua ion. ☐

[1 mark]

30 Which curly arrow shows a movement of electrons that does **not** occur in the mechanism for the reaction between ethanoyl chloride and methanol?

A

B

C

D

[1 mark]

31 Which is a correct statement about the Cl_3^+ ion?

A Its shape is linear.

B It is larger than an I_3^- ion.

C It has more electrons than a Cl_3^- ion.

D It has two bonding pairs and two lone pairs around the central Cl atom.

[1 mark]

32 Which of the following statements about the reactions of chlorine is **not** correct?

A In the presence of sunlight, chlorine reacts with water to form chloride and oxide ions.

B Chlorine water reacts with aqueous potassium iodide to form a brown solution.

C Chlorine reacts with cold dilute aqueous sodium hydroxide to form sodium chloride.

D Chlorine is both oxidised and reduced in its reaction with water.

[1 mark]

Turn over for the next question

Turn over ▶

33 Which structure shows a correct repeating unit of a polymer that could be formed from propane-1,2-diol and propanedioic acid?

A $-O-\overset{\displaystyle O}{\overset{\|}{C}}-CH_2-\overset{\displaystyle O}{\overset{\|}{C}}-O-CH_2-CH_2-CH_2-$ ☐

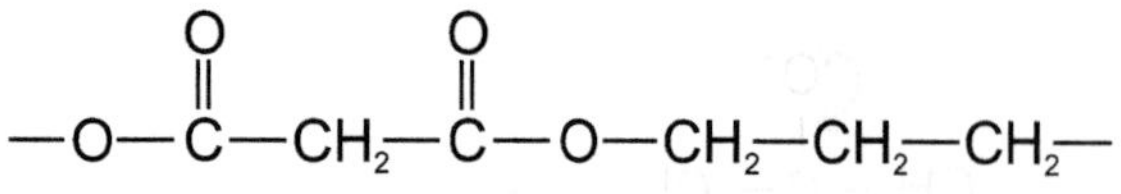

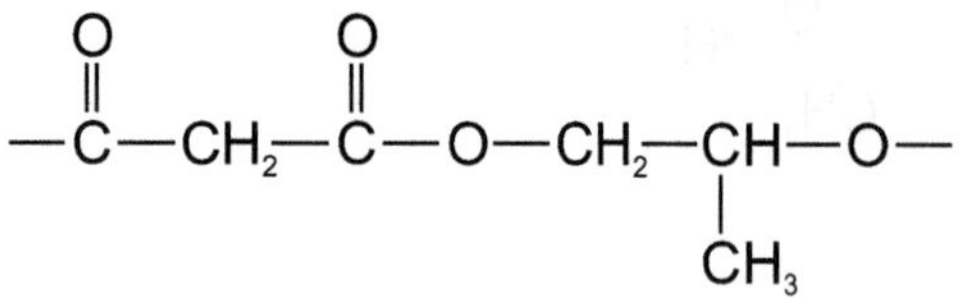

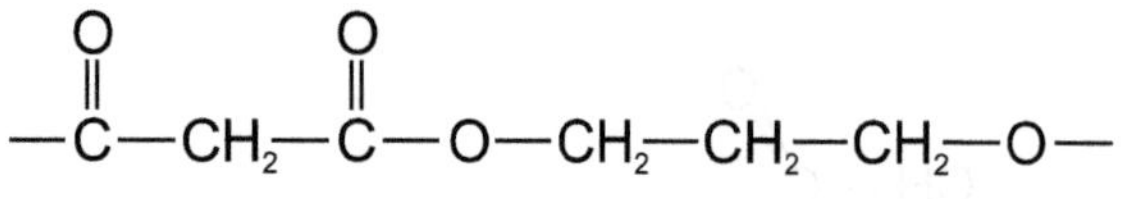

B $-\overset{\displaystyle O}{\overset{\|}{C}}-CH_2-\overset{\displaystyle O}{\overset{\|}{C}}-O-CH_2-\underset{\underset{\displaystyle CH_3}{|}}{CH}-O-$ ☐

C $-\overset{\displaystyle O}{\overset{\|}{C}}-CH_2-\overset{\displaystyle O}{\overset{\|}{C}}-O-CH_2-CH_2-CH_2-O-$ ☐

D $-O-\overset{\displaystyle O}{\overset{\|}{C}}-CH_2-\overset{\displaystyle O}{\overset{\|}{C}}-O-\underset{\underset{\underset{\displaystyle CH_3}{|}}{\underset{\displaystyle CH_2}{|}}}{CH}-$ ☐

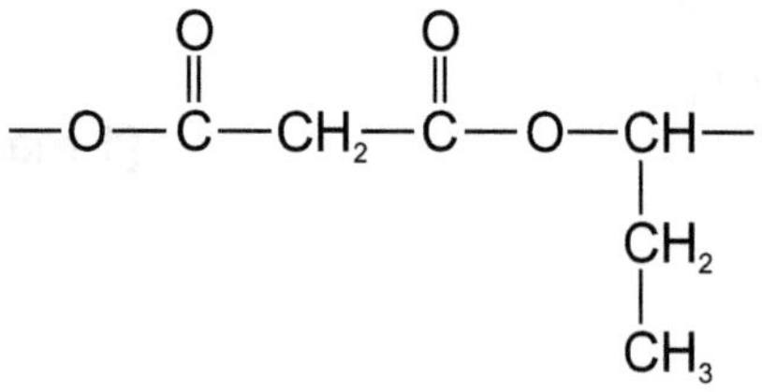

[1 mark]

34 Which statement about P_4O_{10} is **not** correct?

A Phosphorus is in its highest oxidation state. ☐

B It dissolves in water to form hydrogen ions and phosphate ions. ☐

C It has a higher melting point than Na_2O. ☐

D It reacts with bases to form a phosphate salt and water. ☐

[1 mark]

END OF QUESTIONS

A-Level

Chemistry

Exam Board: AQA

Practice Exam Papers
Instructions & Answer Booklet

Exam Set CAP71

<table>
<tr><td>55.8</td></tr>
<tr><td>**Fe**</td></tr>
<tr><td>26</td></tr>
</table>

Fe out your Chemistry exam prep with these CGP Practice Papers!

Just like your average bond enthalpy, AQA A-Level Chemistry can be pretty mean. Fortunately, this brilliant pack from CGP contains a full set of Practice Papers.

They're carefully crafted to be just like the real exams — so you'll know exactly what to expect on the day. We've also included fully worked answers and clear mark schemes for each question. What more could you ask?

(Actually, you could ask for a data booklet. But you don't have to, because we've given you one of those too!)

CGP — still the best! ☺

Our sole aim here at CGP is to produce the highest quality books — carefully written, immaculately presented and dangerously close to being funny*.

Then we work our socks off to get them out to you — at the cheapest possible prices.

*Admittedly these practice papers aren't very funny, since we were too busy concentrating on making them as serious as the real exams. But normally we'd include more jokes, honestly.

Working Out Your Grade

- Complete the full set of papers (Paper 1, Paper 2 and Paper 3).

- Use the answers and mark scheme in this booklet to mark each paper.

- Write down your mark for each paper in the table below.
  Paper 1 and Paper 2 are each worth a maximum of 105 marks.
  Paper 3 is worth a maximum of 90 marks.

- Find your total for the whole exam (out of a maximum of 300 marks)
  by adding up your marks from all three papers.

- Follow the instructions below to estimate your grade.

	Paper 1	Paper 2	Paper 3	Total	Grade
Mark					

Estimating Your Grade

- If you want to get a **rough idea** of the grade you're working at, we suggest you compare the **total mark** you got to the latest set of grade boundaries.

- Grade boundaries are set for each individual exam, so they're likely to **change** from year to year. You can find the latest set of grade boundaries by going to **www.cgpbooks.co.uk/alevelgradeboundaries**

- Jot down the marks required for each grade in the table below so you don't have to refer back to the website. Use these marks to **estimate your grade**.
  If you're borderline, don't push yourself up a grade — the real examiners won't.

Total mark required for each grade						
Grade	A*	A	B	C	D	E
Total mark out of 300						

- Remember, this will only be a **rough guide**, and grade boundaries will be different for different exams, but it should help you to see how you're getting on.

Published by CGP

Editors: Emma Clayton, George Wright and Mary Falkner.

Contributors: Louise Watkins, Megan Pollard and Sarah Binns.

With thanks to Katie Fernandez, Jamie Sinclair and Sarah Pattison for the proofreading.

With thanks to Jan Greenway for the copyright research.

Illustrations by: Sandy Gardner Artist, email sandy@sandygardner.co.uk

Printed by Elanders Ltd, Newcastle upon Tyne.

ISBN: 978 1 78908 464 1

Text, design, layout and original illustrations © Coordination Group Publications Ltd. (CGP) 2020 All rights reserved.

Answers

Practice Paper 1 — Inorganic and Physical Chemistry

1.1 Shape: octahedral *[1 mark]*
Bond angle: 90° *[1 mark]*
An AlF_6^{3-} ion has six pairs of bonding electrons around the central Al atom and no lone pairs *[1 mark]*. The bonding electron pairs all repel each other equally to be as far apart as possible *[1 mark]*.

You're told to use Valence Shell Electron Pair Repulsion Theory to get the answer here. Al is in Group 3 so has three electrons in its outer shell. It has six bonds, so add on another six electrons, plus three extra electrons from the 3− charge, so there are 3 + 6 + 3 = 12 electrons, or 6 pairs, around the Al centre. This matches the number of bonds, so there are no lone pairs.

1.2 Fluorine is a non-polar molecule, so only has van der Waals forces acting between its molecules *[1 mark]*, whereas hydrogen fluoride is polar, so also has permanent dipole interactions, and hydrogen bonding, acting between its molecules *[1 mark]*. These forces are stronger than van der Waals forces, therefore more energy is needed to overcome the intermolecular forces in hydrogen fluoride *[1 mark]*.

1.3 dative covalent/co-ordinate bond *[1 mark]*
A lone pair of electrons on each fluoride ion is donated to the manganese ion *[1 mark]*.

1.4 Potassium fluoride conducts electricity when in solution but not when solid *[1 mark]*. Solid potassium fluoride has a giant ionic lattice structure with strong ionic bonds between oppositely charged ions *[1 mark]*. This means the ions are fixed in position, so cannot move and conduct an electric current / carry an electric charge *[1 mark]*. However, when the ionic compound is dissolved in solution the ions are free to move and conduct an electric current / carry an electric charge *[1 mark]*.

2.1 $Zn + Cu^{2+} \rightarrow Cu + Zn^{2+}$ *[1 mark]*

2.2 Plot the results on a graph with time on the x-axis and temperature on the y-axis *[1 mark]*. Draw two separate lines of best fit — one for the temperature over the first three minutes and one for the temperature from five to ten minutes *[1 mark]*. Extrapolate both lines of best fit back/forwards to the fourth minute *[1 mark]*. The difference in temperature between the two extrapolated lines at the fourth minute is the maximum temperature change for the reaction *[1 mark]*.

2.3 Number of moles of Na_2CO_3 = mass ÷ M_r = 4.10 ÷ 106
$= 0.0386...$ mol
Heat energy transferred to water (q) $= -(\Delta H \times n)$
$= -(-26.5 \times 0.0386...)$
$= 1.025$ kJ
$= 1025$ J
$\Delta T = \dfrac{q}{m \times c} = 1025 \div (50.0 \times 4.18)$
$= 4.9043...$ °C
Maximum temperature $= 21.4 + 4.9043... = 26.3043...$ °C
$= 26.3$ °C (3 s.f.)

[4 marks for the correct answer, otherwise 1 mark for the correct number of moles of Na_2CO_3, 1 mark for the correct value of q, 1 mark for the correct change in temperature.]

It's easy to get your positive and negative signs mixed up when you're answering enthalpy and temperature change questions, so make sure you check your answer seems sensible. The enthalpy change in this question is negative (and you're asked for the 'maximum' temperature), so you know it's an exothermic reaction. That means you should be expecting your answer to be bigger than the starting temperature.

3.1 Relative atomic mass $= ((84 \times 0.56) + (86 \times 9.86) +$
$(87 \times 7.00) + (88 \times 82.58)) \div 100$
$= 87.7$ (1 d.p.)

[2 marks for the correct answer to 1 d.p., otherwise 1 mark for the correct method to work out the relative atomic mass.]

3.2 strontium *[1 mark]*

3.3 mass of ion $= 86 \div (6.022 \times 10^{23} \text{ mol}^{-1}) = 1.428... \times 10^{-22}$ g
$= 1.428... \times 10^{-25}$ kg
KE $= \frac{1}{2} \times 1.428... \times 10^{-25} \times 82200^2 = 4.8247... \times 10^{-16}$ J
$= 4.82 \times 10^{-16}$ J (3 s.f.)

[2 marks for the correct answer, otherwise 1 mark for the correct mass of the ion.]

4.1 SiO_2 has a giant covalent/macromolecular structure with covalent bonds *[1 mark]*, whereas P_4O_{10} has a simple molecular structure with weak intermolecular forces *[1 mark]* and MgO has a giant ionic lattice with ionic bonding between the oppositely charged ions *[1 mark]*. SiO_2 has a higher melting point than P_4O_{10} as the covalent bonds in SiO_2 require more energy to overcome than the weak intermolecular forces between P_4O_{10} molecules *[1 mark]*. SiO_2 has a lower melting point than MgO as the ionic bonds in MgO are stronger and require more energy to overcome than the covalent bonds in SiO_2 *[1 mark]*.

4.2 D *[1 mark]*

4.3 $Al_2O_3 + 6HCl \rightarrow 2AlCl_3 + 3H_2O$ *[1 mark]*

5.1 $Ca_{(g)} \rightarrow Ca^+_{(g)} + e^-$ *[1 mark]*

5.2 Atomic radius increases down Group 2 *[1 mark]*. This is because the number of electron shells increases down the group, so the atoms increase in size *[1 mark]*.

5.3 $Ba_{(s)} + 2H_2O_{(l)} \rightarrow Ba(OH)_{2(aq)} + H_{2(g)}$ *[1 mark]*

5.4 A white precipitate would form *[1 mark]*.

5.5 To remove any other ions that would form a precipitate (such as carbonates or sulfites) *[1 mark]*.

5.6 Sulfuric acid contains sulfate ions, which would form a white precipitate *[1 mark]*.

6.1 oxidising agent *[1 mark]*

6.2 brown *[1 mark]*

6.3 chlorine *[1 mark]*

The oxidation state of chlorine is 0 in Cl_2, +1 in NaClO and −1 in NaCl, so it is both oxidised and reduced in this reaction.

6.4 E.g. HBr is a stronger reducing agent than HCl *[1 mark]*, so the HBr produced in the initial reaction with sulfuric acid will also undergo a reaction with the acid (this time to give off bromine gas, sulfur dioxide gas and water), whereas HCl won't *[1 mark]*. Therefore more acid will be consumed for each mole of NaBr added (compared to NaCl) *[1 mark]*. As more acid is consumed, the concentration of hydrogen ions/acidity will decrease to a greater extent *[1 mark]*.

6.5 How to grade your answer:

Level 0: There is no relevant information. *[No marks]*

Level 1: One stage is covered in detail OR two stages are covered but they are incomplete or contain inaccuracies. The answer is not given in a logical order. *[1 to 2 marks]*

Level 2: Two stages are covered in detail OR all three stages are covered but they are incomplete or contain inaccuracies. The answer is mostly given in a logical order. *[3 to 4 marks]*

Level 3: All three stages are covered in detail and are complete and accurate. The answer is coherent and given in a logical order. *[5 to 6 marks]*

Indicative content:

Stage 1: Addition of acidified silver nitrate solution

Add dilute nitric acid to the solution, followed by a few drops of silver nitrate solution.

The silver ions in the solution will react with any iodide ions present to produce a yellow precipitate of silver iodide.

Formation of a yellow precipitate suggests that iodide ions are present.

Stage 2: Addition of ammonia solution

Adding ammonia solution can confirm the results of the first test.

Add concentrated ammonia solution to the same solution from the first test.

The yellow precipitate will not re-dissolve.

This distinguishes iodide ions from other halides.

Stage 3: Addition of sodium hydroxide solution

Perform this test on a separate sample to the iodide tests.

Add dilute sodium hydroxide solution to the unknown solution.

Then gently heat the mixture.

Hold a damp piece of red litmus paper above the solution.

The reaction of sodium hydroxide with ammonium ions produces ammonia gas and water.

The ammonia gas causes the damp red litmus paper to turn blue.

This confirms the presence of ammonium ions in the solution.

7.1 The yield would decrease *[1 mark]*. There are two moles of gas on the left-hand side of the equation and one mole of gas on the right-hand side *[1 mark]*. Decreasing the pressure causes the equilibrium to shift towards the side with more moles of gas in order to increase the pressure/oppose the decrease in pressure *[1 mark]*.

7.2 Calculate the number of moles of N_2 and H_2 at equilibrium:

From the equation, two moles of NH_3 are produced from one mole of N_2, so 1.4 moles of NH_3 are produced from 0.7 moles of N_2.

Number of moles of N_2 at equilibrium = $1 - 0.7 = 0.3$

1.4 moles of NH_3 are produced from $1.4 \times (3 \div 2) = 2.1$ moles of H_2

Number of moles of H_2 at equilibrium = $3 - 2.1 = 0.9$

Total number of moles = $1.4 + 0.3 + 0.9 = 2.6$

Calculation of mole fractions:

Mole fraction = number of moles ÷ total number of moles

$N_2 = 0.3 \div 2.6 = 0.11538...$

$H_2 = 0.9 \div 2.6 = 0.34615...$

$NH_3 = 1.4 \div 2.6 = 0.53846...$

Calculation of partial pressures:

Partial pressure = mole fraction × total pressure

$p_{N_2} = 0.11538... \times 150 = 17.307...$ kPa

$p_{H_2} = 0.34615... \times 150 = 51.923...$ kPa

$p_{NH_3} = 0.53846... \times 150 = 80.769..$ kPa

Derive K_p expression: $K_p = \dfrac{(p_{NH_3})^2}{p_{N_2} \times (p_{H_2})^3}$

Substitute into K_p expression:

$K_p = 80.769...^2 \div (17.307... \times 51.923...^3)$

$K_p = 2.69 \times 10^{-3}$ (3 s.f.)

[6 marks for the correct answer, otherwise 1 mark for the correct number of moles of N_2 and H_2 at equilibrium, 1 mark for the correct total number of moles, 1 mark for all three mole fractions correct, 1 mark for all three partial pressures correct and 1 mark for the correct expression for K_p.]

Units for K_p = kPa^{-2} *[1 mark]*

Be especially careful to show all your working on questions like this one, no matter how confident you are — there are lots of marks to pick up here even if you don't get the final answer.

7.3 The value of K_p would increase *[1 mark]*. A decrease in temperature causes the equilibrium to shift in favour of the exothermic direction *[1 mark]* to increase the temperature/ oppose the decrease in temperature *[1 mark]*.

8.1 Solutions have a concentration of 1.00 mol dm^{-3}, a temperature of 298 K (25 °C) and a pressure of 100 kPa *[1 mark]*.

8.2 $Zn_{(s)} \mid Zn^{2+}_{(aq)} \parallel Fe^{3+}_{(aq)}, Fe^{2+}_{(aq)} \mid Pt_{(s)}$

[1 mark for left-hand side of salt bridge correct and 1 mark for right-hand side of salt bridge correct.]

Don't forget to include a platinum electrode in your half-cells if they don't have a solid conductive component.

8.3 Oxidation state: +2 *[1 mark]*

Colour of solution: violet/purple *[1 mark]*

The $E^\ominus$ for the VO_2^+/VO^{2+}, VO^{2+}/V^{3+} and V^{3+}/V^{2+} half-cells are all more positive than the $E^\ominus$ of the Zn^{2+}/Zn half-cell *[1 mark]*. Therefore VO_2^+ is reduced to VO^{2+}, VO^{2+} is further reduced to V^{3+} and V^{3+} is further reduced to V^{2+} *[1 mark]*. The $E^\ominus$ for the V^{2+}/V half-cell is less positive than the $E^\ominus$ of the Zn^{2+}/Zn half-cell, therefore V^{2+} cannot be reduced any further *[1 mark]*.

8.4 E.g. fuel cells are more efficient / only produce water as a waste product / do not produce CO_2 *[1 mark]*.

9.1 $K_w = [H^+][OH^-]$

$[H^+] = \dfrac{K_w}{[OH^-]} = \dfrac{1.0 \times 10^{-14}}{0.100} = 1.0 \times 10^{-13}$ mol dm^{-3}

pH $= -\log_{10}[H^+] = -\log_{10}(1.0 \times 10^{-13}) = 13$

[2 marks for the correct answer, otherwise 1 mark for the correct concentration of H^+.]

9.2 pH probe/meter *[1 mark]*

9.3 $K_a = \dfrac{[H^+][CH_3COO^-]}{[CH_3COOH]} \;/\; K_a = \dfrac{[H^+]^2}{[CH_3COOH]}$

[1 mark]

9.4 E.g. when pH = pK_a,

$-\log_{10}[H^+] = -\log_{10}(K_a)$

so $K_a = [H^+]$ *[1 mark]*

From the expression for K_a, in order for K_a to equal $[H^+]$, $[CH_3COO^-] = [CH_3COOH]$ (and this occurs when half of the acid has been neutralised) *[1 mark]*.

The concentrations of the acid and conjugate base are equal at half-neutralisation because every acid molecule that reacts with the base becomes a conjugate base molecule. This means that when the concentration of the acid has fallen by half of its initial value, the concentration of the conjugate base will have risen by the same amount (from 0) to equal it.

9.5 Full neutralisation occurs at 25.0 cm^3

so volume required for half-neutralisation = $25.0 \div 2$

$= 12.5$ cm^3

[1 mark]

Remember, full neutralisation occurs at the equivalence point — on the graph, this is the midpoint of the rapidly-changing section.

9.6 Concentration of HCOOH = $0.025 \div V$

Concentration of HCOONa = $0.0125 \div V$

$K_a = \dfrac{[H^+][HCOO^-]}{[HCOOH]} = \dfrac{[H^+][HCOONa]}{[HCOOH]}$

so $[H^+] = \dfrac{K_a[HCOOH]}{[HCOONa]} = \dfrac{K_a \times (0.025 \div V)}{(0.0125 \div V)} = K_a \times 2$

$= 1.78 \times 10^{-4}$ mol dm$^{-3} \times 2$

$= 3.56 \times 10^{-4}$ mol dm^{-3}

pH $= -\log_{10}(3.56 \times 10^{-4}$ mol dm$^{-3}) = 3.45$ (2 d.p.)

[4 marks for the correct answer, otherwise 1 mark for stating the expression for K_a, 1 mark for rearranging to give an expression for $[H^+]$ and 1 mark for the correct concentration of H^+.]

You can simplify the equation for $[H^+]$ because the volumes on the top and bottom of the fraction are the same, so they cancel out. That just leaves 0.025 ÷ 0.0125, which is 2.

10.1

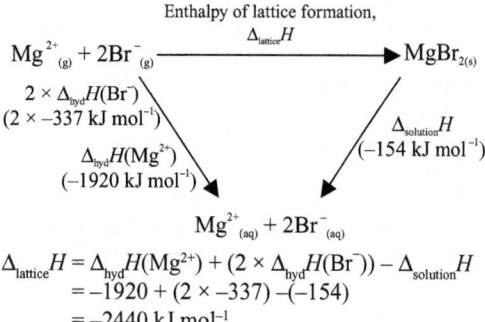

$$\Delta_{lattice}H = \Delta_{hyd}H(Mg^{2+}) + (2 \times \Delta_{hyd}H(Br^-)) - \Delta_{solution}H$$
$$= -1920 + (2 \times -337) - (-154)$$
$$= -2440 \text{ kJ mol}^{-1}$$

[2 marks for the correct answer, otherwise 1 mark for a correct cycle or equation to find the answer.]

Remember to check your standard enthalpy changes carefully. In this question you're asked for the enthalpy of lattice formation — but you could just as easily be asked for the enthalpy of lattice dissociation.

10.2   The values suggest that MgI_2 has some covalent character / the bonds are polarised *[1 mark]*, because the bonding is stronger than in a pure ionic lattice *[1 mark]*.

10.3 $\Delta G = \Delta H - T\Delta S$
The reaction becomes feasible when $\Delta G = 0$, so when
$$T = \frac{\Delta H}{\Delta S}$$
$T = +130 \div (+335 \div 1000)$
$T = 388.05... \text{ K}$
  $= 115\,°C$ (3 s.f.)

[3 marks for the correct answer to three significant figures, otherwise 1 mark for the correct equation for T and 1 mark for the correct temperature in K.]

Remember that ΔS must be converted into kJ mol^{-1} K^{-1}, or ΔH must be converted into J mol^{-1}. To convert the temperature in K to °C, you need to subtract 273 (because 0 °C = 273 K).

11.1   Cr: $1s^2\ 2s^2\ 2p^6\ 3s^2\ 3p^6\ 4s^1\ 3d^5$ /
$1s^2\ 2s^2\ 2p^6\ 3s^2\ 3p^6\ 3d^5\ 4s^1$ *[1 mark]*
Cr^{3+}: $1s^2\ 2s^2\ 2p^6\ 3s^2\ 3p^6\ 3d^3$ *[1 mark]*

11.2   Overall equation:
$MnO_4^- + 8H^+ + 5Fe^{2+} \rightarrow Mn^{2+} + 4H_2O + 5Fe^{3+}$ *[1 mark]*

You can work out equations like this one using half-equations — in this case you'll need the half-equations:
$MnO_4^- + 8H^+ + 5e^- \rightleftharpoons Mn^{2+} + 4H_2O$ and $Fe^{3+} + e^- \rightleftharpoons Fe^{2+}$.
Then you have to multiply the iron half-equation by 5, since it needs to balance out the number of electrons in the manganese half-equation. After that, just combine the two equations and cancel out the electrons — and make sure you get the reactants and products the right way round for each one.

11.3 $[Fe(H_2O)_6]^{3+}_{(aq)} + 3NH_{3(aq)} \rightarrow [Fe(H_2O)_3(OH)_3]_{(s)} + 3NH_4^+{}_{(aq)}$
[1 mark for correct equation and 1 mark for correct state symbols.]

11.4 $[Cu(H_2O)_2(NH_3)_4]^{2+}$ *[1 mark]*
deep blue *[1 mark]*

11.5   If Fe^{2+} was present in solution **X**, the addition of excess sodium hydroxide to the first test tube would produce a green precipitate (which would change from green to brown on standing) *[1 mark]*. When the sodium carbonate solution is added to the second test tube, Fe^{2+} ions would again produce a green precipitate *[1 mark]*.
If Al^{3+} was present, the addition of sodium hydroxide would initially produce a white precipitate, which would then dissolve in excess sodium hydroxide to produce a colourless solution *[1 mark]*. Upon the addition of sodium carbonate to the second sample, any Al^{3+} ions present would produce a white precipitate and bubbling/effervescence *[1 mark]*.

Practice Paper 2 — Organic and Physical Chemistry

1.1   The ethanol is produced by a two-step process:
Photosynthesis: $6CO_2 + 6H_2O \rightarrow C_6H_{12}O_6 + 6O_2$ *[1 mark]*
Fermentation: $C_6H_{12}O_6 \rightarrow 2C_2H_5OH + 2CO_2$ *[1 mark]*
It is consumed by combustion:
Combustion: $2C_2H_5OH + 6O_2 \rightarrow 4CO_2 + 6H_2O$ *[1 mark]*
6 moles of CO_2 are consumed during the photosynthesis step, and in total 6 moles of CO_2 are produced during the fermentation and combustion steps/the same number of moles of CO_2 are consumed as are produced *[1 mark]*.

1.2   E.g. fossil fuels need to be burned to power the machinery to make fertilisers for the crops/harvest the crops/refine and transport the ethanol, which produces CO_2 *[1 mark]*

1.3   E.g. fractional distillation *[1 mark]*

1.4   E.g. mix each alcohol with acidified potassium dichromate and gently heat the mixtures, using distillation apparatus to collect the product *[1 mark]*. Mix the product with Fehling's Solution/Benedict's Solution *[1 mark]*.
The solution that contained pentan-1-ol will produce a brick red precipitate, while nothing would be observed for the pentan-2-ol solution *[1 mark]*.

Heating with the dichromate under distillation conditions will oxidise pentan-1-ol to an aldehyde and pentan-2-ol to a ketone. The test with Fehling's/Benedict's solution can then distinguish between the two. Fehling's and Benedict's solutions contain copper(II) ions which oxidise aldehydes (like the one made by pentan-1-ol) to carboxylic acids, but do not react with ketones such as the one produced from pentan-2-ol. In the reaction with the aldehyde, the copper(II) ions are reduced to form a red copper(I) oxide precipitate.

1.5
$$K_c = \frac{[CH_3OH]}{[CO][H_2]^2} \text{ [1 mark]}$$
units = $mol^{-2}\ dm^6$ *[1 mark]*

1.6   How to grade your answer:

Level 0:	There is no relevant information. *[No marks]*
Level 1:	One stage is covered well OR two stages are covered but they are incomplete and not always accurate. The answer is not in a logical order. *[1 to 2 marks]*
Level 2:	Two stages are covered well OR all three stages are covered but they are incomplete and not always accurate. The answer is mostly in a logical order. *[3 to 4 marks]*
Level 3:	All three stages are covered and are complete and accurate. The answer is coherent and is in a logical order. *[5 to 6 marks]*

Indicative content:
Stage 1: Temperature
The forward reaction is exothermic.
A low temperature causes the equilibrium to favour the forward/exothermic reaction to oppose the decrease in temperature/raise the temperature.
So a low temperature will produce a better yield.
Low temperatures are also cheaper, easier and safer to maintain.
However, low temperatures will give slower reaction rates, so a compromise temperature of 250 °C is used.

Stage 2: Pressure
There are three moles of gas on the left-hand side of the equation and one mole of gas on the right-hand side of the equation.
A high pressure causes the equilibrium to favour the side with fewer moles of gaseous substances, in order to oppose the increase in pressure/reduce the pressure.
A high pressure will therefore cause the equilibrium to favour the right-hand side and will produce a better yield.
High pressures also give a greater reaction rate.
However, high pressures are expensive to create/maintain, so while a high pressure is used, it is limited by cost.

Stage 3: Catalyst
The copper catalyst increases the rate (of both the forward and reverse reactions).
This means that the reactants will be converted into products more quickly.
As the catalyst increases the rate of both the forward and reverse reactions equally, it will have no effect on the yield/ the position of the equilibrium.
However, use of the catalyst allows the reaction to be carried out at a lower temperature without significantly reducing the rate, which helps to improve the yield.
Using a catalyst reduces the energy requirement and therefore reduces the cost of the process.

2.1   Draw a tangent to the curve at time = 0:

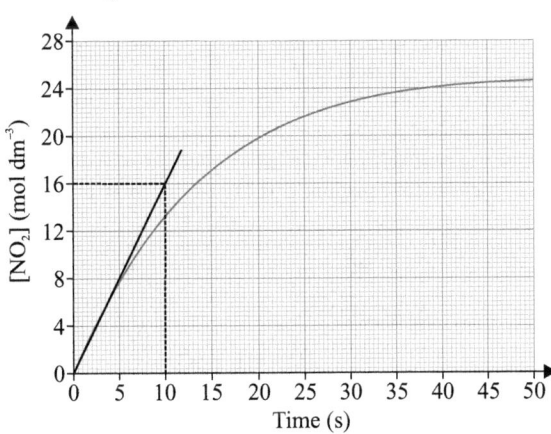

The gradient of the tangent = change in y ÷ change in x
$= 16 \div 10 = 1.6$
initial rate = gradient = 1.6 mol dm⁻³ s⁻¹

$k = \text{rate}/[\text{NO}]^2[\text{O}_2] = 1.6 \div (0.75^2 \times 5.0)$
$= 0.57$

[4 marks for a correct answer in the range 0.53–0.61, otherwise 1 mark for a correct tangent drawn on the graph, 1 mark for calculating the gradient, 1 mark for a correct method to find the rate constant.]
units = dm⁶ mol⁻² s⁻¹ *[1 mark]*

2.2   There would be no effect on the value of the rate constant *[1 mark]*.

2.3   rate = $k[\text{H}_2][\text{NO}]^2$ *[1 mark for each correct reaction order.]*
Comparing experiments 1 and 2, increasing [H₂] by a factor of 2.5 also increases the initial rate by a factor of 2.5, so the reaction is first order with respect to H₂. Comparing experiments 1 and 3, [H₂] and [NO] are both doubled while the initial rate increases by a factor of 8. We would expect the rate to double as a result of doubling [H₂] which means that doubling the concentration of NO must have increased the initial rate by a factor of 4, or 2². The reaction is therefore second order with respect to NO.

2.4 1.6×10^{-2} mol dm⁻³ s⁻¹ *[1 mark]*
To get this answer, you could either scale the initial rate by comparing experiment 4 with one of the previous experiments, or you could use the results of one of the previous experiments to calculate the rate constant and then use that to work out the missing initial rate.

3.1   2-bromo-2-methylpropanoyl chloride *[1 mark]*
3.2   The chlorine and oxygen atoms are more electronegative than carbon so draw electrons away from the carbonyl carbon atom, giving it a slight positive charge *[1 mark]*. This means that this carbon atom is very easily attacked by nucleophiles *[1 mark]*.
3.3   nucleophilic addition-elimination *[1 mark]*

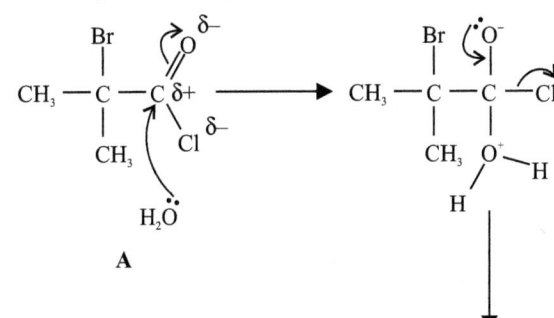

[1 mark for curly arrow on lone pair of O in H₂O to the carbonyl C atom, 1 mark for curly arrow from C=O bond to O, 1 mark for correct intermediate structure including the correct charges on the O atoms, 1 mark for the final three curly arrows.]

3.4   ammonia/NH₃ *[1 mark]*
3.5   three peaks *[1 mark]*
There would be three peaks because there are three different hydrogen environments in this molecule: the six hydrogen atoms in the CH₃ groups attached to the carbon bonded to bromine, the two hydrogen atoms in the CH₂ group after the ester bond and the three hydrogen atoms in the CH₃ group next to the CH₂ group.

3.6   E.g.

[1 mark for any correct isomer with a carboxylic acid group.]
There are several possible isomers you could choose to draw here. The important thing is that it has a carboxylic acid group, as that is what reacts with sodium carbonate.

4.1   Solid ammonium chloride contains covalent bonds between each nitrogen atom and three of the four surrounding hydrogen atoms *[1 mark]*, dative covalent bonds between each nitrogen atom and the fourth hydrogen atom *[1 mark]* and ionic bonds between the ammonium ions and the chloride ions *[1 mark]*.

4.2   moles of NH₃ = moles of NH₄Cl = mass ÷ M_r = 1 ÷ 53.5
$= 0.0186...$

$T = 350 \,°C = (350 + 273) \,K = 623 \,K$
$V = nRT/p = (0.0186... \times 8.31 \times 623) \div (100 \times 10^3)$
$= 9.67... \times 10^{-4} \,m^3$
$= 0.97 \,dm^3$

[4 marks for the correct answer, otherwise 1 mark for the correct number of moles, 1 mark for the correct temperature in K and 1 mark for rearranging the ideal gas equation to find the volume.]
Remember to make sure you're always using the correct units when you're doing a calculation (as well as checking the units of the answer at the end). In this question you have to convert the temperature to K and pressure to Pa before you can use them in the ideal gas equation.

4.3

bond angle: 120° *[1 mark]*

4.4 bonds broken: 4 (S=O) + (O=O)
bonds formed: 6 (S=O)
ΔH = energy of bonds broken – energy of bonds formed
= (4 × 523) + 498 – (6 × 523)
= –548 kJ mol^{-1}
[2 marks for the correct answer, otherwise 1 mark for correct equation to calculate the enthalpy change.]

5.1 compound **H** *[1 mark]*

5.2 E.g. mix samples of both compounds with bromine water and shake *[1 mark]*. Compound **E** would cause the bromine water to change from orange to colourless, whereas nothing would be observed with compound **F** *[1 mark]*.

5.3 Compound **H** would have a peak at 3230-3550 cm^{-1} whereas compound **E** would not *[1 mark]*. Compound **E** would have a peak at 2500-3000 cm^{-1} whereas compound **H** would not *[1 mark]*.
The peak at 3230-3550 cm^{-1} is for the alcohol group in compound H. The peak at 2500-3000 cm^{-1} is for the carboxylic acid group in compound E.

5.4 Compound **F** would be the major product, because compound **F** is formed via a more stable (secondary) carbocation *[1 mark]*.

5.5 Compound **F** has a chiral centre/a carbon atom with four different functional groups *[1 mark]*. Therefore it will exist as two different enantiomers/optical isomers, giving three isomers in total *[1 mark]*. The mixture would have no effect on plane polarised light *[1 mark]* because each enantiomer is produced in equal proportions, and therefore their effects cancel out *[1 mark]*. Compound **G** is achiral, so it also has no effect on plane polarised light *[1 mark]*.

6.1

[1 mark]

[1 mark]

6.2 a strong acid/sulfuric acid *[1 mark]*

6.3 E.g. the dipeptide contains multiple carboxylic acid groups *[1 mark]*, all of which can undergo esterification to give different products *[1 mark]*.

6.4

[1 mark]

7.1 hydrogen bond(s) *[1 mark]*

7.2 Guanine/cytosine form three hydrogen bonds when they join together, whereas adenine/thymine only form two *[1 mark]*. Therefore, more energy is required to separate guanine/cytosine pairs *[1 mark]*.

7.3

[1 mark for the phosphate group correctly bonded to deoxyribose, 1 mark for the deoxyribose correctly bonded to adenine.]

7.4 Cisplatin binds to DNA (causing a kink in the shape of the DNA molecule/creating stronger links between strands) *[1 mark]*, which prevents DNA molecules from unwinding and separating *[1 mark]*.

8.1 $q = mc\Delta T = 200 × 4.18 × 55.2 = 46147.2$ J
moles of propanol = mass ÷ M_r = 1.8 ÷ 60 = 0.03 moles
$\Delta_c H$ = –(46147.2 ÷ 1000) ÷ 0.03 = –1540 kJ mol^{-1} (3 s.f.)
[3 marks for the correct answer, otherwise 1 mark for the correct value of q and 1 mark for the correct number of moles of propanol.]

Remember to include the minus sign when you're calculating the enthalpy change at the end — you know from the question that it's an exothermic reaction, so you should have a negative value for your answer.

8.2 chemical equation: $C_3H_8O + \frac{9}{2}O_2 \rightarrow 3CO_2 + 4H_2O$

$\Delta_c H = \Delta_f H(\text{products}) – \Delta_f H(\text{reactants})$
= (3 × –393.5) + (4 × –285.8) – (–302.7)
= –2021 kJ mol^{-1}
[3 marks for the correct answer, otherwise 1 mark for the correct chemical equation and 1 mark for a correct thermodynamic cycle or method to find the answer.]

8.3 E.g. loss of heat to the surroundings during the experiment *[1 mark]*.

8.4 E.g. increase insulation around the apparatus. / Use a graphical method to find a more accurate temperature change. / Put a lid on the beaker *[1 mark]*.

8.5 Using less water would mean that the temperature change would be higher *[1 mark]*. If the temperature increased over 100 °C the water would start to boil (making it impossible to measure an accurate temperature change) *[1 mark]*.

9.1 **C** *[1 mark]*

Compound J has both an alcohol and a carboxylic acid functional group, so molecules of this compound can react with each other via nucleophilic addition-elimination reactions. This produces an ester/-COOC- linkage between molecules.

9.2

[1 mark]

9.3 E.g. it will cause the chewing gum to be non-biodegradable *[1 mark]*.

*The polymer produced from compound **K** is an addition polymer, therefore its main carbon chain is non-polar and can't be attacked by nucleophiles/easily broken down.*

9.4

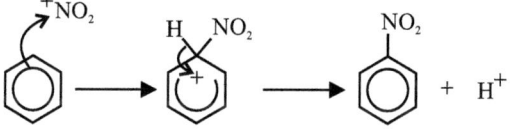

[1 mark]

Extra product = ethanoic acid *[1 mark]*

10.1 Formation of the nitronium ion:
$HNO_3 + H_2SO_4 \rightarrow H_2NO_3^+ + HSO_4^-$,
then $H_2NO_3^+ \rightarrow NO_2^+ + H_2O$
or $HNO_3 + H_2SO_4 \rightarrow NO_2^+ + H_2O + HSO_4^-$
or $HNO_3 + 2H_2SO_4 \rightarrow NO_2^+ + H_3O^+ + 2HSO_4^-$ *[1 mark]*

[3 marks — 1 mark for each curly arrow and 1 mark for correct structure of intermediate.]

10.2 Substitution reactions are favoured over addition reactions because they retain the stable delocalised ring of electrons *[1 mark]*.

10.3 Tin/Sn, concentrated hydrochloric acid and reflux *[1 mark]*, followed by the addition of sodium hydroxide solution *[1 mark]*.

10.4 atom economy
= (M_r of desired product ÷ M_r of reactants) × 100
= ($135 ÷ (93 + 102)$) × 100
= 69.2 %
[2 marks for the correct answer, otherwise 1 mark for all relative molecular masses correct.]

10.5 E.g. reaction B produces HCl as a by-product, which is toxic. / Ethanoic anhydride is cheaper/safer to use than ethanoyl chloride *[1 mark]*.

11 How to grade your answer:
Level 0:   There is no relevant information. *[No marks]*
Level 1:   One stage is covered well OR two stages are covered but they are incomplete and not always accurate. The answer is not in a logical order. *[1 to 2 marks]*
Level 2:   Two stages are covered well OR all three stages are covered but they are incomplete and not always accurate. The answer is mostly in a logical order. *[3 to 5 marks]*
Level 3:   All three stages are covered and are complete and accurate, including determination of the correct structures. The answer is coherent and is in a logical order. *[6 to 8 marks]*
Indicative content:
Stage 1: NMR
The ^{13}C NMR spectrum contains four peaks, so there are four different carbon environments.
This means there must be at least four carbon atoms in the compound.
The peak at ~170 ppm shows that a C=O bond is present, belonging to an ester or carboxylic acid group.
The peaks around 130 ppm show two carbon environments consistent with a C=C bond — more environments would be expected for an aromatic compound.
The final peak at ~55 ppm range corresponds to a carbon atom bonded to a more electronegative atom — either oxygen as an alcohol, ether or ester, or nitrogen as an amine or amide.

Stage 2: Infrared
The peak at ~1700 cm⁻¹ in the infrared spectrum confirms that the compound contains a carbonyl group.
The spectrum has a further peak in this region at ~1600 cm⁻¹ which could confirm the presence of a C=C group.
The lack of a peak in the 2500-3000 cm⁻¹ range suggests that the C=O group is not an acid — therefore by comparison with the NMR data it must be an ester.
There are no peaks above ~3200 cm⁻¹ that could correspond to alcohol or amine groups, therefore the NMR peak at ~55 ppm is confirmed to belong to a carbon atom bonded to the ester oxygen atom.
Stage 3: Formula and structure
The presence of an ester group means that there are at least two oxygen atoms in the molecule along with at least four carbon atoms from the number of NMR peaks.
These atoms would give an m/z of
$(12.000 \times 4) + (15.9990 \times 2) = 79.9980$,
leaving $86.0448 - 79.9980 = 6.0468$.
This difference can only be made up with hydrogen atoms, so there are exactly four carbon and two oxygen atoms in the molecule.
$6.0468/1.0078 = 6$ so the molecule also contains six hydrogen atoms and the molecular formula is $C_4H_6O_2$.
The information about the formula and functional groups present implies the presence of a central C(–O)–O–C ester group within the molecule, as well as an additional C=C group.
This C=C group could go either side of the ester, giving two possible structures.
The two possible structures are:

and

Practice Paper 3

Section A

1.1   They are non-superimposable mirror images *[1 mark]*.

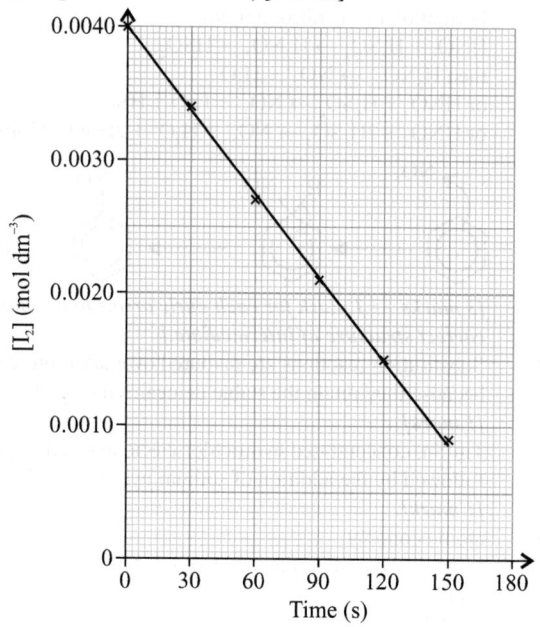

[1 mark]

1.2   2-methylbutanoic acid *[1 mark]*

1.3   Purpose: to neutralise any excess acid *[1 mark]*.
      Precaution: after shaking the mixture, invert the separating funnel and open the tap *[1 mark]*.

This is to allow the carbon dioxide gas formed during the neutralisation to escape, stopping pressure from building up in the funnel.

1.4   drying agent/to remove water *[1 mark]*

1.5   It can be removed by filtration *[1 mark]*.

This works because calcium chloride does not dissolve in the organic layer.

1.6   The student may have predicted this because the alcohol has hydrogen bonds between its molecules, which take a lot of energy to break *[1 mark]*. However, the ester is a larger molecule than the alcohol (with more electrons) and so has stronger van der Waals forces between molecules *[1 mark]*. The ester also has more C–O bonds (which are polar), so has greater permanent dipole-dipole forces *[1 mark]*.

You can tell from Figure 1 that the alcohol used was ethanol, and therefore that the ester is a much bigger molecule.

1.7   E.g. place a suitable collecting vessel at the end of the condenser and heat to between 78 and 133 °C to distil off any alcohol *[1 mark]*. Change the collection vessel when the temperature on the thermometer is just below 133 °C and continue to heat until just below 176 °C to distil off the ester *[1 mark]*. The purity of the sample can be confirmed by measuring its boiling point and comparing the result to the data book value/133 °C *[1 mark]*.

1.8   alcohol = CH_3CH_2OH (ethanol)
      $M_r(CH_3CH_2OH) = 46$
      mass of alcohol in solution = $(8.50 \div 100) \times 78.5 = 6.6725$ g
      number of moles = mass $\div M_r = 6.6725 \div 46$
      $= 0.14505...$ moles
      $= 0.145$ moles (3 s.f.)

      [3 marks for the correct answer, otherwise 1 mark for the correct relative molecular mass of the alcohol (ethanol) and 1 mark for the correct mass of the alcohol.]

1.9   Atom economy
      $= M_r$ of ester $\div$ total M_r of products (or reactants)
      M_r of ester = 130, M_r of water = 18
      Atom economy = $(130 \div 148) \times 100 = 87.8\%$

      [2 marks for the correct answer, otherwise 1 mark for correct relative molecular masses of the ester and water, or the ester and the reactants.]

2.1   substitution *[1 mark]*

In this reaction, an iodine atom simply replaces a hydrogen atom, so it is an example of a substitution reaction.

2.2   So the rate of reaction will only depend on the concentration of I_2. / So that I_2 will be the limiting reagent *[1 mark]*

2.3

Time (s)	Relative Absorbance	$[I_2]$ (mol dm^{-3})
0	0.89	0.0040
30	0.76	0.0034
60	0.60	0.0027
90	0.47	0.0021
120	0.34	0.0015
150	0.20	0.0009

(accept ±0.0001 mol dm^{-3}) *[1 mark]*

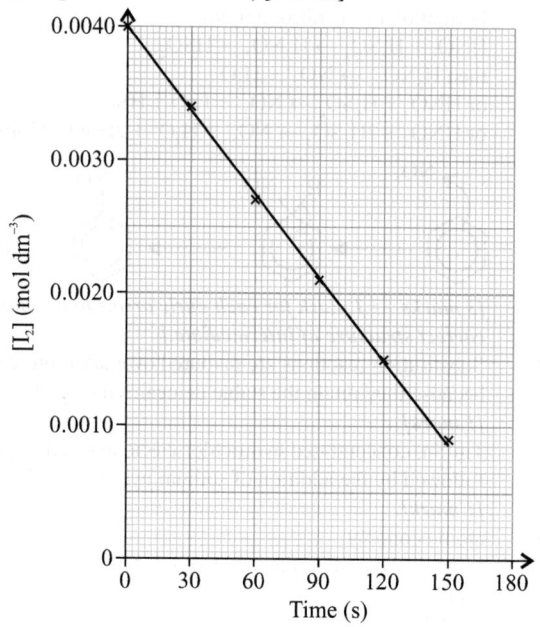

[1 mark for all points plotted correctly and 1 mark for the line of best fit.]

2.4   zero order *[1 mark]*

The line has a constant gradient so the rate does not change with the concentration of iodine.

2.5   rate = $k[H^+][CH_3COCH_3]$ *[1 mark]*
      units of $k = $ mol^{-1} dm^3 s^{-1} *[1 mark]*

The slowest step of a reaction is the rate determining step. In this rate determining step there is one molecule of propanone and one H$^+$ ion so the reaction is first order with respect to each of these components. Iodine does not appear in the rate equation as it is not involved in the rate determining step — you can also see from the graph that the order of the reaction with respect to $[I_2]$ is zero.

2.6   Levelled response:

Level 0:   There is no relevant information. *[No marks]*

Level 1:   One stage is covered well OR two stages are covered but they are incomplete and not always accurate. The answer is not in a logical order. *[1 to 2 marks]*

Level 2:   Two stages are covered well OR all three stages are covered but they are incomplete and not always accurate. The answer is mostly in a logical order. *[3 to 4 marks]*

Level 3:   All three stages are covered and are complete and accurate. The answer is coherent and is in a logical order. *[5 to 6 marks]*

Indicative content:

Stage 1: Method

Carry out the reaction several times using different initial concentrations of NO.

Keep the concentration of Cl_2 constant and in excess.

Start timing the reaction as soon as the reactants are mixed.

Measure the concentration of NO at regular time intervals.

Stage 2: Determination of initial rate
Plot a graph of concentration of NO (*y*) against time (*x*).
Draw a line of best fit (which will be a curve on the graph).
Draw a tangent to the curve at time = 0.
Find the gradient of the tangent to work out the initial rate.
Repeat for all initial concentrations of NO.
Stage 3: Determination of order
Record each initial concentration of NO and its corresponding initial rate.
Compare the data for two of the experiment runs.
E.g. if the concentration of NO doubles and the initial rate increases by a factor of four between the two runs, the reaction is second order with respect to NO as $4 = 2^2$.
Confirm this result using different pairs of experiment runs.

3.1   E.g. to prevent multiple substitutions occurring. / To reduce the amount of secondary/tertiary amines formed. / Ammonia vapour would escape through the top of the condenser *[1 mark]*.

3.2

$$CH_3 - \underset{\underset{CH_3}{|}}{\overset{\overset{CH_3}{|}}{C}} - \underset{\overset{|}{CH_3}}{\overset{H}{N}}$$ *[1 mark]*

3.3   Compound **A** has an alkyl group, which is electron releasing/has a positive inductive effect/pushes electrons onto the amine group *[1 mark]*. This increases the availability of the lone pair on the nitrogen atom to accept a proton *[1 mark]*.

3.4   The hydroxide ion can act as a nucleophile/react via nucleophilic substitution (to produce compound **B**) *[1 mark]*, or it can act as a base/react via an elimination reaction (to produce compounds **C** and **D**) *[1 mark]*.

*Compound **B** is an alcohol, formed by the substitution of the bromine atom by the hydroxide. Compounds **C** and **D** are alkenes, formed by the elimination of HBr from the halogenoalkane.*

3.5   Like Figure 4, the IR spectrum of a pure sample of compound **B** would contain an O–H alcohol absorption peak at 3230-3550 cm^{-1} *[1 mark]*. The IR spectrum of pure compound **B** would not contain the peaks in the C=C/ carbon-carbon double bond region at 1620-1680 cm^{-1} which are present in Figure 4 *[1 mark]*. The fingerprint region/ region below 1500 cm^{-1} of the IR spectrum for a pure sample of compound **B** would not match the spectrum in Figure 4 *[1 mark]*.

3.6   The student could add acidified potassium dichromate to the sample and heat the mixture *[1 mark]*. An orange to green colour change would confirm the presence of an alcohol/ compound **B** *[1 mark]*.

4.1 $K_a = \dfrac{[H^+][[Al(H_2O)_5(OH)]^{2+}]}{[[Al(H_2O)_6]^{3+}]}$ *[1 mark]*

4.2 $[H^+] = 10^{-pH} = 10^{-2.81} = 1.5488... \times 10^{-3}$ mol dm^{-3}
$K_a = 10^{-pK_a} = 10^{-4.85} = 1.4125... \times 10^{-5}$ mol dm^{-3}
Assume that $[[Al(H_2O)_5(OH)]^{2+}] = [H^+]$, so that
$[[Al(H_2O)_6]^{3+}] = [H^+]^2 \div K_a$
              $= (1.5488... \times 10^{-3})^2 \div 1.4125... \times 10^{-5}$
              $= 0.170$ mol dm^{-3} (3 s.f.)
[3 marks for the correct answer, otherwise 1 mark for the correct values of $[H^+]$ and K_a and 1 mark for a correct equation to find the concentration.]

4.3   moles of $AlCl_3 = 0.170 \times (20.0 \div 1000) = 3.40 \times 10^{-3}$ mol
moles of $AgCl = 3 \times 3.40 \times 10^{-3} = 0.0102$ mol
theoretical mass of $AgCl$ = number of moles × M_r
                      $= 0.0102 \times (107.9 + 35.5)$
                      $= 1.46268$ g
so actual mass $= 0.32 \times 1.46268 = 0.468$ g
[3 marks for the correct answer, otherwise 1 mark for the correct number of moles of silver chloride and 1 mark for the correct theoretical mass of silver chloride.]
If you used the concentration of 0.191 mol dm^{-3} given in the question, you should get a mass of 0.526 g.

4.4   white *[1 mark]*
Fluted filter paper has a large surface area, which allows the solution to travel quickly through the filter paper *[1 mark]*.

4.5   To remove any soluble impurities *[1 mark]* and traces of water *[1 mark]*.

4.6   number of moles = concentration × volume
                    $= 0.0150 \times (250 \div 1000)$
                    $= 3.75 \times 10^{-3}$ moles
mass = number of moles × $M_r = 3.75 \times 10^{-3} \times 342.3 = 1.28$ g
[2 marks for the correct answer, otherwise 1 mark for the correct number of moles.]

4.7   Any two from: e.g. weigh the solid by difference / include washings when transferring from one container to another / rinse the stirring rod into the solution / invert the flask several times to ensure all the solid dissolves.
[2 marks — 1 mark for each correct suggestion.]

4.8   A white precipitate and bubbles of gas would be observed *[1 mark]*. The products are $[Al(H_2O)_3(OH)_3]$ (the white precipitate) and CO_2 (the gas) *[1 mark]*.

Section B

5   A *[1 mark]*
In general, larger molecules have higher boiling points due to stronger van der Waals forces, but the presence of other types of intermolecular force can change this order. Hydrogen fluoride is smaller than the other hydrogen halides, however it has extensive hydrogen bonding, therefore its intermolecular forces are much stronger and its boiling point is higher.

6   C *[1 mark]*
The greater the difference between the experimental and theoretical lattice enthalpies, the greater the covalent character.

7   D *[1 mark]*

8   C *[1 mark]*

9   B *[1 mark]*
$q = mc\Delta T = 40 \times 4.18 \times 16.2 = 2708.64$ J = 2710 J (3 s.f.)

10   D *[1 mark]*
Both sides of the double bond must have two different groups for there to be stereoisomerism.

11   B *[1 mark]*

12   D *[1 mark]*
An EMF of +0.32 V means that the cell must be made up of the Zn^{2+}/Zn and Fe^{2+}/Fe half-cells. These both have solid components (the metals themselves), so don't require an inert electrode.

13   C *[1 mark]*
You can work this out by writing out the balanced equation:
$C_4H_{10} + 6.5O_2 \rightarrow 4CO_2 + 5H_2O$

14   A *[1 mark]*
Sample weight = 5.244 − 4.798 = 0.446 g
Total uncertainty = ±0.0010 g, as there are two weighings with an uncertainty of ±0.0005 g each.
Percentage uncertainty = (0.0010 ÷ 0.446) × 100 = 0.224%

15   B *[1 mark]*

16   D *[1 mark]*

17 B [1 mark]

The formula of barium hydroxide is $Ba(OH)_2$, so each $Ba(OH)_2$ produces two OH^- ions.

So, $[OH^-] = 0.38 \times 2 = 0.76 \text{ mol dm}^{-3}$

$[H^+] = K_w \div [OH^-] = 1.0 \times 10^{-14} \div 0.76 = 1.315... \times 10^{-14} \text{ mol dm}^{-3}$

$pH = -\log_{10}[H^+] = -\log_{10}[1.315... \times 10^{-14}] = 13.9$

18 C [1 mark]

Number of moles of $NaClO = 4.20 \div 74.5 = 0.05638...$ per 100 g

So there are $0.05638...$ moles per 90.1 cm^3.

$[ClO^-] = [NaClO] = 0.05638... \div (90.1 \div 1000) = 0.63 \text{ mol dm}^{-3}$

19 D [1 mark]

Compound D has three peaks, compound B has four peaks and compounds A and C each have five peaks in their ^{13}C NMR spectra.

20 B [1 mark]

The A_r of magnesium is closer to 24 than 25, so the most abundant isotope and hence the tallest peak in the mass spectrum must be ^{24}Mg.

21 B [1 mark]

22 A [1 mark]

$\Delta S = 192 - 95.6 = 96.4 \text{ J K}^{-1} \text{ mol}^{-1}$

The reaction becomes feasible when $T = \Delta H / \Delta S$.

$T = (23.4 \times 1000) \div 96.4 = 242.738... \text{ K} = -30.261... °C = -30.3 °C$

23 A [1 mark]

24 D [1 mark]

$n = pV \div RT = (101000 \times 373 \times 10^{-6}) \div (8.31 \times 403) = 0.0112 \text{ moles}$

$M_r = 0.897 \div 0.0112 = 79.7$

25 A [1 mark]

26 A [1 mark]

The average of the concordant titres (2 and 4) is 15.35 cm^3 — don't forget to subtract the initial burette volumes.

number of moles of MnO_4^- = concentration × volume

$= 0.02 \times (15.35 \div 1000)$

$= 3.07 \times 10^{-4}$

number of moles of $Fe^{2+} = 3.07 \times 10^{-4} \times 5 = 1.535 \times 10^{-3}$

mass = number of moles × $A_r = 1.535 \times 10^{-3} \times 55.8 = 0.0857 \text{ g}$

27 B [1 mark]

mole fraction of **X** = partial pressure of **X** ÷ total pressure

$= 16.2 \div 70.0 = 0.231...$

total moles of gas = moles of **X** ÷ mole fraction of **X**

$= 0.350 \div 0.231... = 1.51$

28 C [1 mark]

For vanillin to exist as a pair of enantiomers, it would need to have a carbon atom with four different groups bonded to it.

29 C [1 mark]

30 D [1 mark]

31 D [1 mark]

This ion must have a central Cl atom with the remaining two Cl atoms bonded to it. Chlorine is in Group 7 so the central Cl atom has seven valence electrons. Adding one electron for each bond and subtracting one for the positive charge gives $7 + 2 - 1 = 8$ electrons. Four of these make up the two bonding pairs, so there are two lone pairs around the central Cl atom as well.

32 A [1 mark]

33 B [1 mark]

34 C [1 mark]

A-Level

Chemistry

Exam Board: AQA

Practice Exam Papers

Data Booklet

Spectroscopic Data

^{1}H NMR chemical shift data

Type of proton	δ / ppm
ROH	0.5 - 5.0
RCH$_3$	0.7 - 1.2
RNH$_2$	1.0 - 4.5
R$_2$CH$_2$	1.2 - 1.4
R$_3$CH	1.4 - 1.6
R—C(=O)—C—H	2.1 - 2.6
R—O—C—H	3.1 - 3.9
RCH$_2$Cl or RCH$_2$Br	3.1 - 4.2
R—C(=O)—O—C—H	3.7 - 4.1
R,H C=C	4.5 - 6.0
R—C(=O)—H	9.0 - 10.0
R—C(=O)—O—H	10.0 - 12.0

^{13}C NMR chemical shift data

Type of carbon	δ / ppm
—C—C—	5 - 40
R—C—Cl or R—C—Br	10 - 70
R—C(=O)—C—	20 - 50
R—C—N	25 - 60
—C—O— (alcohols, ethers or esters)	50 - 90
C=C	90 - 150
R—C≡N	110 - 125
⬡	110 - 160
R—C—O (esters or acids)	160 - 185
R—C—O (aldehydes or ketones)	190 - 220

Infrared absorption data

Bond	Wavenumber / cm^{-1}
C–C	750 - 1100
C–O	1000 - 1300
C=C	1620 - 1680
C=O	1680 - 1750
C≡N	2220 - 2260
O–H (acids)	2500 - 3000
C–H	2850 - 3300
O–H (alcohols)	3230 - 3550
N–H (amines)	3300 - 3500

Organic Structures

Bases

adenine	thymine	guanine	cytosine

Amino acids

alanine	cysteine	serine	aspartic acid

$H_2N-CH-COOH$ with CH_3 (alanine)

$H_2N-CH-COOH$ with CH_2-SH (cysteine)

$H_2N-CH-COOH$ with CH_2-OH (serine)

$H_2N-CH-COOH$ with CH_2-COOH (aspartic acid)

phenylalanine: $H_2N-CH-COOH$ with CH_2 and phenyl ring

lysine: $H_2N-CH-COOH$ with $CH_2-CH_2-CH_2-CH_2-NH_2$

Sugars

glucose, 2-deoxyribose

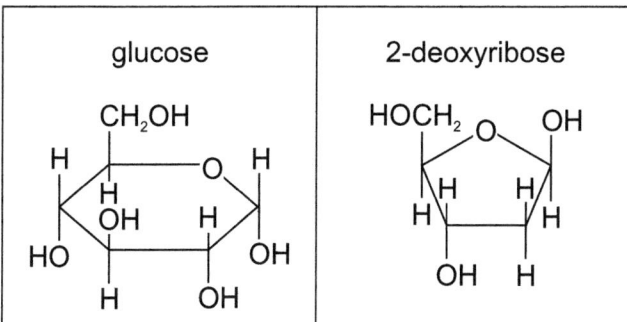

Phosphate

$^-O-P(=O)(OH)(OH)$

Haem B

The Periodic Table

relative atomic mass →

atomic (proton) number →

1.0 H Hydrogen 1

1	2												3	4	5	6	7	0
																		4.0 He helium 2
6.9 Li lithium 3	9.0 Be beryllium 4												10.8 B boron 5	12.0 C carbon 6	14.0 N nitrogen 7	16.0 O oxygen 8	19.0 F fluorine 9	20.2 Ne neon 10
23.0 Na sodium 11	24.3 Mg magnesium 12												27.0 Al aluminium 13	28.1 Si silicon 14	31.0 P phosphorus 15	32.1 S sulfur 16	35.5 Cl chlorine 17	39.9 Ar argon 18
39.1 K potassium 19	40.1 Ca calcium 20	45.0 Sc scandium 21	47.9 Ti titanium 22	50.9 V vanadium 23	52.0 Cr chromium 24	54.9 Mn manganese 25	55.8 Fe iron 26	58.9 Co cobalt 27	58.7 Ni nickel 28	63.5 Cu copper 29	65.4 Zn zinc 30		69.7 Ga gallium 31	72.6 Ge germanium 32	74.9 As arsenic 33	79.0 Se selenium 34	79.9 Br bromine 35	83.8 Kr krypton 36
85.5 Rb rubidium 37	87.6 Sr strontium 38	88.9 Y yttrium 39	91.2 Zr zirconium 40	92.9 Nb niobium 41	96.0 Mo molybdenum 42	[97] Tc technetium 43	101.1 Ru ruthenium 44	102.9 Rh rhodium 45	106.4 Pd palladium 46	107.9 Ag silver 47	112.4 Cd cadmium 48		114.8 In indium 49	118.7 Sn tin 50	121.8 Sb antimony 51	127.6 Te tellurium 52	126.9 I iodine 53	131.3 Xe xenon 54
132.9 Cs caesium 55	137.3 Ba barium 56	138.9 La* lanthanum 57	178.5 Hf hafnium 72	180.9 Ta tantalum 73	183.8 W tungsten 74	186.2 Re rhenium 75	190.2 Os osmium 76	192.2 Ir iridium 77	195.1 Pt platinum 78	197.0 Au gold 79	200.6 Hg mercury 80		204.4 Tl thallium 81	207.2 Pb lead 82	209.0 Bi bismuth 83	[209] Po polonium 84	[210] At astatine 85	[222] Rn radon 86
[223] Fr francium 87	[226] Ra radium 88	[227] Ac† actinium 89	[267] Rf rutherfordium 104	[270] Db dubnium 105	[269] Sg seaborgium 106	[270] Bh bohrium 107	[270] Hs hassium 108	[278] Mt meitnerium 109	[281] Ds darmstadtium 110	[281] Rg roentgenium 111	[285] Cn copernicium 112		[286] Nh nihonium 113	[289] Fl flerovium 114	[289] Mc moscovium 115	[293] Lv livermorium 116	[294] Ts tennessine 117	[294] Og oganesson 118

* 58 – 71 Lanthanides

† 90 – 103 Actinides

140.1 Ce cerium 58	140.9 Pr praseodymium 59	144.2 Nd neodymium 60	[145] Pm promethium 61	150.4 Sm samarium 62	152.0 Eu europium 63	157.3 Gd gadolinium 64	158.9 Tb terbium 65	162.5 Dy dysprosium 66	164.9 Ho holmium 67	167.3 Er erbium 68	168.9 Tm thulium 69	173.0 Yb ytterbium 70	175.0 Lu lutetium 71
232.0 Th thorium 90	231.0 Pa protactinium 91	238.0 U uranium 92	[237] Np neptunium 93	[244] Pu plutonium 94	[243] Am americium 95	[247] Cm curium 96	[247] Bk berkelium 97	[251] Cf californium 98	[252] Es einsteinium 99	[257] Fm fermium 100	[258] Md mendelevium 101	[259] No nobelium 102	[262] Lr lawrencium 103

CAP71U